Crimes Against Nature

Crimes Against Nature

Illegal Industries and the Global Environment

DONALD R. LIDDICK

PRAEGER

AN IMPRINT OF ABC-CLIO, LLC
Santa Barbara, California • Denver, Colorado • Oxford, England

Library of Congress Cataloging-in-Publication Data

Liddick, Don.
   Crimes against nature : illegal industries and the global environment / Donald R. Liddick.
      p. cm.
   Includes bibliographical references and index.
   ISBN 978–0–313–38464–6 (hbk. : alk. paper) — ISBN 978–0–313–38465–3 (ebook)
   1. Offenses against the environment. 2. Industries—Environmental aspects. I. Title.
   HV6401.L53   2011
   363.7—dc22   2010043755

ISBN: 978–0–313–38464–6
EISBN: 978–0–313–38465–3

15 14 13 12 11   1 2 3 4 5

This book is also available on the World Wide Web as an eBook.
Visit www.abc-clio.com for details.

Praeger
An Imprint of ABC-CLIO, LLC

ABC-CLIO, LLC
130 Cremona Drive, P.O. Box 1911
Santa Barbara, California 93116-1911

This book is printed on acid-free paper ∞

Manufactured in the United States of America

I would like to dedicate this book to my academic mentors,
Alan A. Block and Richard A. Ball

The Earth is defiled by its people; they have disobeyed the laws,
violated the statutes and broken the everlasting covenant.
Isaiah, chapter 24, verse 5 (*New International Version*–NIV)

Contents

CHAPTER 1

The Nature of the Problem

Perhaps the greatest challenge of the twenty-first century will be balancing human consumption and needs with the sound stewardship of our physical environment. Fortunately, human prosperity and environmental health are not mutually exclusive. The application of private capital, the production of wealth, the advancement of the human condition, and environmental protection are objectives that can and should be pursued with equal vigor, especially as they are endeavors that complement each other. Yet the impediments to attaining these objectives are legion. Identifying and establishing the proper balance will be difficult, and subject to ideological arguments and political fights over definitions of economic, social, and environmental justice. Moreover, population growth, rapid technological advances, the dynamics of global politics, and the liberalization of trade will continue to exacerbate the difficulties in realizing the correct formula for human and environmental advancement. This arduous task is made even more problematic by unscrupulous private elites and corrupt public officials who exploit consumer demand and misapply capital in a manner that depletes natural resources with little regard for sustainability. Subsequent overconsumption unnecessarily drains the natural world of species and materials, and wrecks social, economic, and environmental havoc.

The goal of this book is not to identify or define in precise terms the proper balance between human and environmental concerns, but to describe and analyze illicit and quasi-licit behaviors that confound the search for that balance. Crimes that negatively impact the environment may be categorized in one of two ways: those that cater to the ever-growing problem of garbage production, especially hazardous wastes; and the illegal harvesting or extraction of natural resources. Human

population growth and rapid technological advancement has precipitated a crisis in waste management, where the expense of responsible
disposal has bred an illicit multi-billion-dollar global traffic in garbage
and hazardous materials. Consumer demand likewise drives black markets in a range of environmental products, living and dead. The illegal
traffic in wildlife, including endangered species, may rival the illicit
drug trade in size and profitability, while the overharvesting of fish
and timber has precipitated severe environmental degradation around
the globe. The scope of illegal entrepreneurship is broad and variable,
and includes highly sophisticated organized crime groups, otherwise-
legitimate corporations and corporate officers, corrupt public officials
at the local, national, and international levels, and informal networks
of individuals that may be localized and ad hoc, or international and
characterized by longevity.

The following anecdotes may serve as an introduction to the problem, and exemplify the broad range of criminal behaviors of interest:

- Located in an area of converging currents between California and
  Hawaii, the Pacific Ocean "garbage patch" is a swirling vortex of
  plastic waste estimated to be twice the size of the state of Texas.
  Large pieces of discarded waste break down into billions of
  microscopic plastic particles that are ingested by marine life and
  passed up the food chain. The waste is thought to derive primarily from discarded plastics that wash down storm sewers and
  rivers from the U.S. Pacific Coast and Japan. In 2010, researchers
  reported on the existence of an Atlantic garbage patch between
  Bermuda and the Azores. The National Oceanic and Atmos-
  pheric Administration estimates that as many as 100,000 marine
  mammals are killed each year as a consequence of floating trash
  in the world's oceans.[1]

- Large amounts of hazardous waste come from decommissioned
  ships. Every year, 600–700 large oceangoing vessels are taken
  out of service and towed to scrapyards in Asia that do not comply
  with international norms. The French military aircraft carrier
  Clemenceau, which contains 100 tons of asbestos, was to be
  shipped to India for dismantling, but the move was halted when
  environmentalists successfully argued that the asbestos could
  not be safely disposed of in the designated port where safety rules
  for workers are negligible.[2]

- In 2009, 1,200 tons of British garbage—including used syringes,
  dirty diapers, and toilet seats—sat rotting in two Brazilian ports
  after having arrived on container ships. The trash, some of which

included computer components (electronic components, or "e-waste," contains numerous toxic substances), was destined for Brazilian companies that said they were expecting recyclable plastic.[3]

- In December 2002, 37 beheaded turtles washed up on beaches in the Bazaruto Archipelago National Park in Mozambique. Then in 2003, the World Wildlife Fund (WWF) reported 40 more turtles either beheaded or with their throats cut washed up in the same area. Authorities suspected longline vessels illegally fishing for sharks in the restricted area had snagged the turtles accidentally and then discarded them (called "by-catch," a problem in which numerous marine species not specifically targeted are nevertheless killed by commercial trawlers). Mozambican authorities sent in soldiers and eventually apprehended an illegal longliner, but only after an exchange of gunfire and the launching of a rocket-propelled grenade.[4]

- In June 2009, the Algerian Coast Guard seized a Turkish fishing vessel, two tugboats, and an Algerian ship off the coast of eastern Algeria. The Turkish boat had a huge net containing 210 tons of live red tuna, which had been illegally transferred from the Algerian trawler. In 2010, an Algerian court sentenced nine men to three years in prison, fined the ship owners $108 million, and seized the vessels. Among the convicted persons were the central director of fishing in Algeria and the secretary general of the Algerian Fisheries Ministry.[5]

- Since the fall of the Soviet Union, poaching of sturgeon in and around the Caspian Sea has become rampant. Caviar smuggled by organized crime gangs can sell for as much as $2,250 in Europe. The Union for the Conservation of Nature (IUCN) has said that 17 of 27 sturgeon species are critically endangered, and that 85% of all sturgeon is at risk of extinction.[6]

- In Italy, about 158 organized crime "families" traffic 35 million tons of garbage annually. In recent years, the criminals have diversified into moving and dumping toxic waste. Italian dairy farmers accept payments for having toxins dumped in their fields, hazardous waste is mixed with other products like cement that is used to construct homes, and all manner of toxic chemicals— even radioactive waste—is simply dumped in landfills, rivers, and the Mediterranean Sea. Industrial and public officials work in concert with so-called "eco-mafias," resulting in cases where millions of tons of toxic waste simply "disappears." In Caserta (in Naples), where the Camorra controls the waste trade, the local population experienced a 400% increase in the cancer rate over a recent five-year period.[7]

- A luxury store owner in Bangkok was arrested in 2007 for illegally importing and selling shahtoosh wool shawls. Shahtoosh is derived from a critically endangered Tibetan antelope. The case exemplifies the feckless deterrent value of Thailand's wildlife laws, as the store owner was sentenced to two years in prison, but eventually received only a short term on probation and a $300 fine. The shawls were worth at least $20,000.[8]

- In South Africa, poaching of rhinoceros has escalated to a 15-year high. In March 2010, rangers discovered a rhino struggling for life in the Kruger National Park—the animal had been tranquilized and its horn removed, leaving a gaping wound (the rhino was subsequently euthanized). Authorities note that international crime syndicates that have entered the rhino-horn traffic use helicopters and military-grade weapons to poach. The black market traffic in rhino horn is fueled by the trade in traditional Asian medicines and the economic boom in Southeast and East Asia.[9]

- In 2010, three Vietnamese government officials were arrested on suspicion of being involved in the transport of illegal timber into Koh Kong province. Since Prime Minister Hun Sen warned military officials that he would no longer tolerate illegal logging, there have been numerous seizures of illegally harvested timber. However, public officials convicted of timber trafficking in Vietnam typically receive light sentences and avoid prison terms.[10]

- After Hamid Karzai banned domestic logging in Afghanistan in 2006, insurgent Taliban forces in Kunar Province developed a black market trade in timber. The Taliban smuggles illegal logs into Pakistan along the same routes they use to move troops and weapons. Revenue from the illicit trade is used to finance the war against U.S. forces in Afghanistan.[11]

- The Tanjung Putting conservation park on Borneo, home to one of Indonesia's last surviving orangutan colonies, has been decimated by illegal logging—10.5 million cubic feet of logs is illegally harvested there annually. Official corruption is central to illicit logging in the archipelago. In 2001, Chinese-owned ships off Borneo transporting $3 million worth of illegal timber was seized—officials believe that the logs belonged to companies associated with Abdul Rasyid, who was elected to Indonesia's Supreme Parliament in 1999.[12]

- In Peru, illegal loggers raid the Alto Purus National Park to harvest a valuable species of mahogany. In 2003, government agents charged with inspecting logs transported downstream had their camp destroyed by arson.[13] Elsewhere in South America, Amazon deforestation was up 4% in the first half of 2008—that year, a group of about 3,000 people, angered over a crackdown on illegal logging, attacked a government office and environmental workers in the Amazonian city of Paragominas.[14]

At the most fundamental level, the crimes examined in this book arise from the regulation and, in some cases, the prohibition of highly demanded goods and services. The utility or fecklessness of particular rules aside, demand for relatively inexpensive waste disposal services, wildlife and wildlife products, fish, and timber and wood products in a highly regulated marketplace creates lucrative black markets. A range of actors willingly exploit this demand with little or no regard for environmental, social, and economic consequences. The illegal income generated by these various black markets is detailed in Table 1.1.

In addition to the inherent value of the natural world and its denizens for their own sake, concern for the severe environmental damage caused by these illegal industries implicates human health and prosperity across a number of dimensions. Overharvesting, the decimation of species, improper disposal of hazardous wastes, and loss in biodiversity will negatively impact humans not only by making the planet less habitable, but also by eliminating or reducing natural resources that, when used prudently, advance the human condition—the development of medicines being one prominent example.

Table 1.1. Illicit Markets for Environmental Goods and Services with Estimated Annual Values

Illegal Market	Annual Illicit Revenue (estimated)
Wildlife Smuggling	$20 billion
Illegal Fishing	$16.5 billion
Illegal Logging	$15 billion
Illegal Garbage/Hazardous Waste Trafficking	$11 billion

Source: http://www.havocscope.com/activities/environmental-goods (accessed October 4, 2010).

Illegal trafficking in hazardous waste, timber, fish, and wildlife causes economic harm and social disruption on top of the despoliation of the natural world. Illegal industries distort the legitimate marketplace and undermine businesses that choose to play by the rules, and consumers are denied the opportunity to make more responsible choices. Developing nations are robbed of their natural resources, and governments denied revenues that might be used to benefit their citizens. Profits are so great they are used to finance conflict and war, and public officials and the very entities established to police these economic sectors are systematically corrupted. Moreover, legitimate capital is applied in such a way that governments are obliged to make decisions that tend to facilitate illegal practices and maximize profits for a few elites at the expense of the environment and impoverished human populations.

MAP FOR THE BOOK

This work will explore four criminal sectors that adversely impact the global environment: the illicit traffic in garbage and hazardous wastes; the illegal traffic in wildlife; illegal, unreported, and unregulated (IUU) fishing; and illegal logging. The examination of each topic will include a description of the crimes committed and the parties involved, an analysis of the factors that give rise to and perpetuate the criminal networks observed, an assessment of the environmental, economic, and social impacts, and the range of established and potential responses to the problem. Because the criminal activities to be examined are not confined to any one country or region, the scope of the study will be global. A brief overview of each topic/chapter may be helpful at this juncture.

Chapter 2 details the illicit traffic in garbage and hazardous wastes. Approximately 500 million tons of hazardous waste is produced globally each year, with perhaps 40 million tons trafficked across international borders—a significant portion of this traffic is illicit. In recent decades, growing concern for the environment and the growth in garbage and waste produced has led to more rigid laws governing disposal. But this has also significantly increased the cost of safe disposal, providing an opportunity for illegal entrepreneurs to step in and offer less expensive waste management services (illegal dumping). Transnational organized crime groups, corporate polluters, public officials, and networks of people in both private and public spheres reap huge profits from the traffic in household garbage, toxic chemicals, electronic waste (e-waste), ozone-depleting substances (ODS), plastics, and even radioactive waste. Toxic substances and garbage are often transported around the globe, usually from wealthier nations to countries in the Third World, where governments hungry for revenue accept the waste for a fee, but then fail to process and dispose of it safely.[15]

In Chapter 3, the illegal trade in wildlife is examined. The traffic is global in scope, and propelled by huge demand for specimens living and dead, animal parts, and products derived from them. Many trafficked specimens are nearly extinct or critically endangered species. Tens of thousands of primates and millions of birds, reptiles, cacti, tropical fish, plants, and invertebrates are removed from the wild every year to be used for clothing, food, entertainment, scientific experiments, and traditional medicines. Some exotic species are taken merely for the pet market, or to satisfy specialist collectors.[16] Specimens can be exceptionally valuable and therefore attract a range of criminals who develop sophisticated methods for harvesting, concealment, and transport. The international trade is characterized by flexible distribution lines and networks of intermediaries, from harvesters through middlemen and on to wholesalers, exporters, processors, and retailers.[17] The consequences of wildlife trafficking include the decimation of some species, a reduction in biodiversity, the spread of disease, and the corruption of public officials. The World Wildlife Fund (WWF) estimates that the profits from illegal wildlife trafficking range from $15 billion to $25 billion per year,[18] and may rival drug and arms trafficking in its significance as a criminal enterprise.[19]

The problem of illegal, unreported, and unregulated (IUU) fishing is described and analyzed in Chapter 4. IUU fishing may account for as much as a third of fish catches worldwide, contributing to the more general problem of overharvesting—the United Nations Food and Agriculture Organization (FAO) has concluded that 75% of the world's fish stocks are fully exploited, overexploited, or depleted. Alongside the problem of an overcapacity of legitimate fishing trawlers in the world fleet, illicit fishing practices have precipitated the collapse or near collapse of some fish stocks.[20] IUU fishing also produces significant negative impacts on broader marine ecosystems, and damages the food security and livelihood of coastal populations in developing countries—the fish stocks along the coasts of Africa have been especially hard hit. IUU fishing may cost developing countries as much as $15 billion annually.[21]

Chapter 5 explores the practice of illegal logging and the traffic in illegally sourced forest products. Illegal logging is a transnational crime problem that may account for as much as 10% of the global timber trade. In addition to severe environmental damage associated with deforestation, the illicit trade in timber compromises international security and is frequently linked to violent conflicts, human rights violations, money laundering, and organized crime.[22] Over 12 million hectares of forest disappeared annually during the 1990s, while approximately 95% of West Africa's forests are already gone.[23] Elsewhere, forests have been nearly erased in parts of Eastern Europe and the Russian Far East. Likewise, large portions of Central America, the Amazon basin, Southeast Asia, and the Indonesian archipelago have suffered substantial losses of forest cover, due in large part to illegal logging.[24] The illicit harvesting and traffic of timber is fueled by demand in Europe, the United States, China, and Japan, and is facilitated through official corruption, violence, and ineffectual governance. The bulky nature of the commodity ensures that the illicit sector of the timber industry involves the active participation of large timber corporations, as well as the collusion of military, police, and government officials.[25]

A WORD ON SOURCES AND APPROACH

The balance between the use of natural resources—or for that matter, even framing wildlife, fish, and trees as "resources" for human exploitation—and the protection of the natural world is inherently political. One may find relatively extreme viewpoints on either side of the debate, as well as thoughtful and well-supported arguments. Many of the sources used in this book are derived from environmental NGOs, who quite naturally take a position geared toward conservation and less toward consumer and business interests. Therefore, it is very important that all sources used here are assessed for accuracy by asking relevant questions:

- Who are the authors?
- Do the authors have a specific agenda? What is it?
- Are there manifest biases in the source?
- What is the tone of the source? (Are any opinions expressed overbearing? Do recommendations comport with the facts?)
- What methods of data collection and analysis were used? Are the methods sound?

All sources used for this book were viewed with appropriate skepticism. Unsatisfactory answers to the above questions sometimes resulted in a source being discounted. However, the presence of specific recommendations and/or opinions in the sources did not usually translate into unsupportable bias. *Opinions* and recommendations in the sources (and those of the present author) included in the narrative of this work are recognizable as such, and always, in the presentation of information, every effort is made to delineate opinion and interpretation from observable facts. The purpose here is to describe and analyze the criminal enterprises of interest, and to suggest some possible responses.

NOTES

1. "Underwater Plastic Waste Threatens World's Food Chain," redOrbit, March 27, 2008, http://www.redorbit.com/news/science/1314953/underwater_plastic_waste_threatens_worlds_food_chain/index.html (accessed October 7, 2010); "2nd Garbage Patch Discovered," *Tribune Review*, April 16, 2010, A3.

2. Julio Godoy, "Environment-France: Dismantling End-of-Life Ships Requires Global Answers," Inter Press Service, June 19, 2006, http://www.ipsnews.net/news.asp?idnews=33675 (accessed September 17, 2010).

3. "Britain's Filthy Garbage Causes Stink in Ports," *Tribune Review*, July 17, 2009, A2.

4. *Study and Analysis of the Status of IUU Fishing in the SADC Region and an Estimate of the Economic, Social and Biological Impacts, Volume 2, Main Report* (London: Marine Resources Assessment Group Ltd., 2008).

5. "Four Algerians, Five Turks Jailed for Illegal Fishing," Illegal-Fishing.info, April 5, 2010, http://www.illegal-fishing.info/item_single.php?item=news&item=news&item_id=4641&approach_id= (accessed September 17, 2010).

6. "Caviar Hunters Pushing Sturgeon to Extinction's Edge," Illegal-Fishing.info, March 18, 2010, http://www.illegal-fishing.info/item_single.php?item=news&item_id=4624&approach_id= (accessed September 17, 2010).

7. Francesca Colombo, "Mafia Dominates Garbage Industry," Tierramérica, 2003, http://www.tierramerica.net/2003/0623/iarticulo.shtml (accessed September 17, 2010).

8. Sarah Janicke, "Luxury Store Owner Convicted for Wildlife Trafficking," World Wildlife Fund, August 27, 2007, http://www.worldwildlife.org/who/media/press/2007/WWFPresitem987.html (accessed September 17, 2010).

9. "Rhino Poaching Soars," News24.com, March 21, 2010, http://www.news24.com/SciTech/News/Rhino-poaching-soars-20100321 (accessed September 17, 2010).

10. "Government Officials Arrested in Logging Bust," Illegal-Logging.info, April 5, 2010, http://www.illegal-logging.info/item_single.php?it_id=4321&it=news (accessed September 17, 2010).

11. "Illegal Logging Financing Taliban Attacks on U.S. Troops," Illegal-Logging.info, April 16, 2010, http://www.illegal-logging.info/item_single.php?it_id=4353&it=news (accessed September 17, 2010).

12. Simon Montlake, "Indonesia Battles Illegal Timber Trade," Special to the *Christian Science Monitor*, February 27, 2002, http://www.csmonitor.com/2002/0227/p07s01-woap.html (accessed September 17, 2010).

13. Chris Fagan and Diego Shoobridge, *The Race for Peru's Last Mahogany Trees: Illegal Logging and the Alto Purus National Park* (Salt Lake City, UT: Round River Conservation Studies, 2007).

14. "Violent Mob Objects to Crackdown on Illegal Logging," *Tribune Review*, November 25, 2008, A2.

15. Jennifer Clapp, "The Illicit Trade in Hazardous Wastes and CFCs: International Responses to Environmental Bads," *Trends in Organized Crime* 3, no. 2 (1997): 14–18; Christoph Hilz, *The International Toxic Waste Trade* (New York: Van Nostrand Reinhold, 1992); Bill D. Moyers, *The Global Dumping Ground* (Santa Ana, CA: Seven Locks Press, 1990); C. A. Anyinam, "Transboundary Movements of Hazardous Wastes: The Case of Toxic Waste Dumping in Africa," *International Journal of Health Services* 21, no. 4 (1991): 759–77; Laura A. Strohm, "The Environmental Politics of the International Waste Trade," *Journal of Environment and Development* 2, no. 2 (Summer 1993): 129–53; Donald J. Rebovich, *Dangerous Ground: The World of Hazardous Waste Crime* (New Brunswick, NJ: Transaction Publishers, 1992).

16. Jane Holden, *By Hook or by Crook: A Reference Manual on Illegal Wildlife Trade and Prosecutions in the United Kingdom* (TRAFFIC International, 1998).

17. Francesco Colombo, "Animal Trafficking," CommonDreams.org, September 6, 2003, http://www.commondreams.org/headlines03/0906-06.htm (accessed September 17, 2010); Dee Cook, Martin Roberts, and Jason Lowther, *The International Wildlife Trade and Organised Crime: A Review of the Evidence and the Role of the UK* (Regional Research Institute, University of Wolverhampton, 2002).

18. World Wildlife Fund, http://www.worldwildlife.org (accessed September 17, 2010); Havoscope, http://havocscope.com/ranking/products.

19. Havoscope, http://havoscope.com/ranking/products.

20. *Review of the Impacts of Illegal, Unreported and Unregulated Fishing on Developing Countries, Final Report* (London: Marine Resources Assessment Group Ltd., 2002); *Pirates and Profiteers: How Pirate Fishing Fleets Are Robbing People and Oceans* (London: Environmental Justice Foundation, 2005).

21. *Pirates and Profiteers.*

22. James Hewitt, *Failing the Forests: Europe's Illegal Timber Trade*, http://assets.panda.org/downloads/failingforests.pdf (Surrey, UK: WWF-UK, 2005); Rob Glastra, *Cut and Run: Illegal Logging and Timber Trade in the Tropics* (Ottawa: IDRC Books, 2005).

23. *Illegal Logging, Governance, and Trade: 2005 Joint NGO Conference*, http://www.fern.org/media/documents/document_1650_1659.pdf (FERN, Greenpeace, WWF, 2005).

24. Holden, *By Hook or by Crook*.

25. Duncan Brack, Kevin Gray, and Gavin Hayman, *Controlling the International Trade in Illegally Logged Timber and Wood Products*, A study prepared for the UK Department for International Development (London: Royal Institute of International Affairs, 2002); *Illegal Logging, Governance and Trade*, 2005.

CHAPTER 2

The Traffic in Garbage and Hazardous Wastes

Modern societies produce a huge amount of trash, from household garbage to more hazardous substances that are the by-products of industry. Governmental and nongovernmental entities recognized and acted on the growing problem in the latter half of the twentieth century, but as environmental regulations tighten and the costs of legal disposal increase, opportunities for illicit traffickers in hazardous wastes continue to expand. A broad range of actors have learned how to profit from garbage, including highly organized transnational criminals, corporate polluters, corrupt public officials, and informal networks of individuals in the public and private sectors. The traffic in household garbage, toxic chemicals, electronic waste, ozone-depleting substances (ODS), plastics, and even radioactive waste may rival the international drug trade in scope and profitability. The negative environmental and social consequences from this illicit traffic are extreme.

DIMENSIONS OF THE PROBLEM

Approximately 400–500 million tons of hazardous waste is produced globally each year, with 35–40 million tons crossing international borders—a significant portion of this traffic is illicit. Beginning in the 1980s, globalization and the liberalization of international trade policies made it easier for legitimate entities to conduct business, but it also facilitated the growth of transnational organized crime—all manner of goods, including waste materials, could now more easily cross international borders. More strict environmental laws and regulations concurrently precipitated a steep rise in the costs of safe and legal disposal, and so it was that the perfect storm of conditions created the opportunity for the illicit (and cheaper) dumping of nonhazardous and toxic wastes.[1]

With savings from illegal disposal ranging from 200% to 300% in the Netherlands and 400% in Italy, the economic incentive for evading domestic and international environmental regulations becomes clear.[2] In fact, the environmental and social harms produced by the illicit disposal of waste are often ignored because the activity presents economic benefits for both developed countries and the Third World—the former cut disposal costs while the latter welcome the revenue the waste imports generate. An example of the North-South price differential illustrates the engine that drives the trade: in the 1980s, the disposal of toxic waste in the United States cost about $250 per ton, but in Africa, the price was only $40.[3] The consequence of differential pricing, regulations, and wages is that the flow of garbage typically goes from relatively rich to poorer nations: from North to South, and from the United States and Western Europe to Eastern Europe, Asia, and Africa. Lack of legislation, poor enforcement, and light penalties all make the risks associated with illicit waste trafficking minimal and worthwhile.[4]

While illicit disposal has clear domestic implications, it is also an international crime problem that generates approximately $10–12 billion per year, including $1–2 billion for established organized crime groups like the 'Ndrangheta, the Camorra, the Yakuza, and Israeli gangs.[5] And while criminal entities profit, illegal waste shipments affect the economic viability of lawful businesses that *do* comply with environmental regulations.[6]

The environmental and social consequences of waste trafficking are greatest on Earth's poorer and developing nations. Africa became a favorite dumping ground by the late 1980s due to weak environmental laws, low wages, and ineffective controls over customs officials. Many African countries plagued with war, famine, and poverty were in desperate need of foreign exchange, and so welcomed much of the waste revenue. Unfortunately, many of the shipments were poorly contained and contaminated groundwater supplies and the soil.[7] Some infamous examples of waste dumping in Africa include:

> Polychlorinated biphenyls (PCBs) waste was shipped to Koko, Nigeria, in 1988—a farmer rented his land for $100 a month and was told 8,000 leaking barrels were filled with fertilizer. The barrels burst in the heat and toxic waste sickened local residents.[8]
>
> Fifteen thousand tons of toxic incinerator ash from Philadelphia was dumped in Guinea by a Norwegian waste firm. The waste was mislabeled as raw material for building bricks, but eventually the Norwegian consul general and four additional government

officials were implicated in the scheme. The ash was eventually shipped back to the United States, where it was disposed of in a landfill.[9]

The government of Guinea-Bissau was offered four times the value of its GNP if it would accept 15 million tons of toxic wastes. The offer was originally accepted, but other countries forced Guinea-Bissau to decline.[10]

In 1991, in the midst of war and famine, Somalia received a waste disposal proposal that was originally accepted by the health minister of the deposed government. The official was allegedly offered a large bribe for accepting the contract, and although the deal was believed to have been canceled, several European waste trading firms agreed to pay the Somali government $80 million to take 500,000 metric tons of waste over 20 years (the firms stood to make $8–10 million per shipment).[11]

By the early 1990s, there were 50 documented cases of toxic waste dumping in Africa affecting over half of all African nations. In recent years, the shipment of electronic waste (e-waste) to Nigeria has become an expanding problem. Five hundred containers of used computers come into Lagos, Nigeria, each month, imported mostly from North America and Europe to fuel the growing electronics market in Africa. While much of the imports are secondhand goods intended for repair and resale, a significant portion is simply hazardous waste material exported and imported with the explicit intent of cheap (and unsafe) disposal. Most of the equipment is not tested for functionality prior to export-import, so it is impossible to determine which materials are "goods" and what simply amounts to waste (it is estimated that anywhere from 25% to 75% is simply waste material containing hazardous elements). Clearly, international and domestic environmental laws are violated. The problem is that in Lagos, almost all unrecoverable e-waste is disposed of improperly by dumping into unlined, unmonitored landfills, often close to groundwater sources, or else set afire, with toxins released into the air.[12]

In 1991, the United States exported 15 tons of banned pesticides per day into various nations (many in Africa) that had no ban. Some of these toxins were given away under the guise of "aid packages." However, the shipments typically contained more pesticides than the recipients needed, and much of the materials exceeded their "use by" dates. Some countries import pesticides and later enact domestic bans, creating stockpiles and backlogs. For example, in Sudan, there are substantial stockpiles of DDT from the 1960s, despite the country's having banned the chemical in 1980.[13]

Elsewhere, the importation of trash, hazardous and nonhazardous, has become a major problem for China. China exports 16 billion tons of goods to the United Kingdom annually, and receives 1.9 million tons of garbage in return.[14] Much of the additional waste imported to China comes from the United States, Europe, Japan, and South Korea.[15] In 2007, China's State Environmental Protection Agency began an investigation into British dumping of garbage in Guangdong province, where the long coastline and proximity to Macao and Hong Kong make the region a prime smuggling area. Authorities there are kept busy fighting against the black market in old car parts, computer components, and discarded household appliances, much of it imported from overseas and sold cheap on the mainland where there is heavy demand for inexpensive appliances. In one instance, the customs authority in Guangzhou intercepted 236.9 tons of smuggled trash in just 35 days. In another case, 119 pieces of vacuum cleaners, electric irons, stereo components, and other electric appliances were discovered underneath a small vessel at the mouth of the Zhujiang River.[16] In 2001, Guangzhou customs intercepted 2,326 tires, 8,414 pieces from old domestic electric appliances, 339 computers, and 84 used cars. In 2002, 466 tons of smuggled household garbage was sent back to Japan from China's Taizhou port—it had been hidden under 1,200 tons of scrap metal. Dutch authorities seized 1,600 tons of waste in 2005. The cargo was officially declared as recovered paper on its way from the United Kingdom to China, but it actually contained bales of compacted household waste, food packaging and residues, plastic bags, and waste wood and textiles. The waste was transported by truck and ferry to Dutch ports, where the bales were transferred to the shipping containers. None of the three countries had given permission, and the waste was eventually shipped back to the United Kingdom.[17]

In no place is illicit trafficking in garbage and hazardous wastes more evident than in Italy, where significant organized crime groups like the Cosa Nostra, the Camorra, and the 'Ndrangheta exercise control over the trade. In addition to household garbage, waste trafficking in Italy involves a broad range of materials such as "dust from smoke abatement in iron and metal industries, incinerator ashes, sludge from water treatment processes in the chemical industry, acid sludge, sludge from tanneries, transformers containing contaminated oil, de-oiled earth, and miscellaneous waste made up of plastic."[18] When not exported, hazardous wastes are mixed with other materials to make bricks or resurface roads, used as raw materials to make fertilizer (which subsequently transfers chrome, cadmium, lead, and nickel up the food chain), and simply dumped on the land and in the Mediterranean Sea.[19]

Authorities believe $8.8 billion a year is earned from all environmental crimes in Italy.[20]

The Camorra dominates the garbage industry in Naples, and at one time controlled the entire waste disposal cycle. In 1994, Italy appointed a special commission to pry away waste disposal from Camorra-run companies, with only limited success. These mob companies routinely undercut legitimate operations and win contracts from local authorities—dumps are filled with household trash and mixed with industrial waste trucked in from around Italy, or else sold as a toxic fertilizer blend.[21]

In just one year, Italian officials said that 11 million metric tons of toxic and industrial waste was disposed of in some 2,000 illegal dumps, or else it was diverted to local waterways and the Mediterranean. In 1997, authorities documented 53 separate Italian organized crime groups that were trafficking in hazardous wastes—not just dumping it domestically, but shipping it to illegal sites in Albania, Eastern Europe, and the African West Coast. A year 2000 International Crime Threat Assessment concluded that about half of the 80 million metric tons of waste produced annually in Italy "disappears."[22] Some of the waste dumped is radioactive—police recently seized 10,000 tons of wood pellets contaminated with caesium-137 that had been imported from Lithuania.[23] In another case, a mafia turncoat admitted to blowing up and sinking a ship off the Calabrian coast that was carrying 120 barrels of radioactive waste—the Cunsky was only one of 32 vessels hauling toxic waste that had been sunk by the Mafia in the Mediterranean.[24] Moreover, the problem of waste disposal in Italy remains largely unaddressed, as in late 2007, dumps were full, garbage collections ceased, and piles of stinking garbage grew to mountains in the city of Naples.[25] Various factors aggravate the problem, including low levels of public awareness concerning the harm caused by eco-crimes, delays in proper regulation, strong territorial control by organized crime groups, poor business ethics, the influence of industrial lobbying firms that work to avoid the high costs of safe disposal, and corruption at high levels of government.[26]

THE ROLE OF ORGANIZED CRIME

In some regions and in some sectors of the waste industry, well-established organized crime groups play a significant role in illicit trafficking and disposal. In addition to the prominent case of Italy (discussed above), the illegal disposal of hazardous wastes by organized crime groups has been well documented in the United States—in New York and New Jersey, organized crime dominated garbage hauling for decades.[27] At times, American Cosa Nostra groups or affiliates have exerted monopolistic control over the private sanitation industry through the infiltration of labor unions, the manipulation of employer trade associations, and the incorporation of their own hazardous waste disposal firms. Crime groups also control or own landfills that accept illicit hazardous waste shipments. Moreover, organized crime's profits from illegal dumping are enhanced through public corruption and collusion with private industry and businesses.[28] In one infamous case, major industrial corporations signed over 270,000 gallons of liquid chemical waste to a firm that simply did not exist.[29]

Municipal garbage and hazardous wastes were not legally distinguishable in the United States until the Resource Conservation and Recovery Act (RCRA) of 1976. The law established procedures for classifying hazardous substances, mandated the creation of a manifest system that would document the movement of hazardous wastes from their generation to their safe disposal, and authorized the states to register corporate waste generators and license hauling and disposal firms.[30] Naturally, the RCRA immediately created a huge demand for hazardous waste hauling and disposal services—as a consequence, contravention of the Act became commonplace, and the entry of organized crime elements into the new hazardous waste trade was readily established. The RCRA was poorly implemented and enforced from the start—the lack of a legitimate hazardous waste industry at that time necessitated interim licensing and bred lax monitoring of the manifest system. In fact, the manipulation of manifests allowed corporate entities to "orphan" their waste, and thus escape liability. Even minus cases of public corruption and regulatory incompetence, private waste generators effectively lobbied Congress so that the RCRA would demand less of them, minimize their liability, and ultimately make the industry amenable (though perhaps not purposefully) to organized crime infiltration.[31]

The economic factors that facilitate organized crime's involvement in waste disposal are readily discernible. Perhaps the most significant element is price inelasticity: an increase in price does not equally reduce demand for the service. "Waste" is also an ambiguous product, the nature of which is sometimes easily concealed or manipulated. In addition, numerous factors determine the pricing of waste disposal services, so that customers have difficulty determining what a "fair" market price might be—this opens the door for infiltration by organized crime elements. Even when customers do successfully seek out lower prices among competitors not owned or controlled by organized crime groups, unfair competition from those "mobbed-up" firms may induce otherwise-legitimate collectors to bribe landfill operators to falsify documents indicating waste had been disposed of at the land-fill site, when in actuality it had been simply dumped elsewhere.[32] Organized criminals (and those driven to criminality by monopolistic or oligopolistic markets) have commingled hazardous waste with ordi-nary garbage (a 20-cubic-yard dumpster full of dry garbage can absorb sixty 55-gallon drums of hazardous liquid waste), released liquid wastes onto city streets and into sewers, concealed it in sludge and dumped it on the land, mixed flammable hazardous wastes with fuel oil to be sold as pure heating oil, and sprayed toxic waste onto rural roads to control dust.[33]

Yet waste trafficking is not perpetrated only by established and well-organized crime groups. Research demonstrates that a wide range of societal players are involved, and includes "conspiracies between waste producers, collection and transport companies, storage firms, manag-ers of dump sites, chemists, specialized laboratories, and even farm-ers."[34] One study that looked at waste hauling and disposal in Maine, Maryland, New Jersey, and Pennsylvania found organized crime involvement in only 3 of 71 case studies—most of the identified crimi-nals were organized crime "associates," as opposed to actual members of Mafia families. The INTERPOL Pollution Crime Working Group found through 35 case studies that organized crime was involved in a variety of pollution crimes, including the illegal import/export of waste, illegal hazardous waste disposal, and the illegal movement of ozone-depleting substances—yet the criminal enterprises observed were not highly organized or structured and tended to be informal, coming together and disbanding as opportunities arose.[35] Moreover, with the notable exceptions of Italy and Ireland, in the European Union, the participation of organized crime in environmental crimes appears to be fairly uncommon.[36] In general, it seems that criminal organization in waste trafficking may be best characterized as ad hoc, where generators, haulers, treatment specialists, storage providers, and disposal players simply agree to violate regulations to save money and increase profits. Perhaps the structure of criminal enterprises in waste trafficking consists simply of informal networks and working relationships meant to exploit opportunities as they arise. Still, the pre-cise nature of criminal organization in environmental crime generally and waste trafficking in particular remains unclear, and it is certainly possible that more highly structured crime groups will be attracted to this lucrative illicit marketplace.

In one sector of organized crime entrepreneurship, the dumping of toxic wastes is incidental to the principal activity. Drug trafficking is largely based on the cultivation and processing of illicit crops such as the opium poppy, the coca plant, and marijuana—practices that have serious environmental consequences, including soil pollution, water pollution, and deforestation. Refiners of heroin and cocaine dump toxic chemicals and other waste by-products into streams and rivers; or else the waste is buried, contaminating the soil and groundwater sources. In the process of maceration and washing coca leaf to make coca paste, lime, gasoline, sulfuric acid, kerosene, ammonia sodium carbonate, and potassium carbonate are routinely dumped on the land and in rivers. Every year in Colombia, approximately 20 million liters of chemical by-products end up in the headwaters of the Orinoco and Amazon rivers. In the Huallaga Basin, fish are almost nonexistent, and many of those that remain are not edible. In addition to exterminating or mutating entire species, agro-chemicals decrease the quality of potable water and present a substantial health threat to indigenous populations. Chemical wastes reduce oxygen in the water, alter water pH levels, and ultimately poison plants and fish. Illicit labs also produce explosions, causing damage to both the environment and humans.[37]

Slash-and-burn techniques for clearing land to grow coca and opium contribute to deforestation and soil erosion. In Peru, increased coca cultivation in the Upper Huallaga Valley is responsible for the stripping of one million hectares of tropical forest resources. Coca is best grown in highlands and the sloping areas of forests high in alkaloid—such areas provide better drainage and a secluded setting, but also are highly susceptible to soil erosion. Again, the land is usually cleared by slash-and-burn techniques, which leaves neither remaining vegetation nor a mechanism for soil replenishment. Coca is harvested three to four times annually, so the subsequent defoliation exposes the soil to wind erosion as well. Sediment and muddy water from deforestation also blocks sunlight necessary for aquatic plants, while compounds in fertilizer produce too much algae—as a consequence, many Amazon River tributaries are nearly devoid of plant and animal life.[38]

Widespread marijuana cultivation also causes environmental harm due to the use of herbicides and pesticides. In the United States, national forests and parks are popular with Mexican marijuana growers. Typically, banned weed and insect killers are smuggled to marijuana farms, and plant-growth hormones are dumped into streams. In some cases, streams are diverted for miles in PVC pipes. Growers sprinkle rat poison to deter wildlife, and deer and bear poaching is common around marijuana grow sites. Areas negatively impacted by illicit marijuana farming include California, the Cascade Mountains, eastern Kentucky, Tennessee, and West Virginia. Seven grow sites discovered on U.S. forest land in California in 2007 and 2008 covered 1,800 square miles of the Sequoia National Forest.

Money for eradication is in the U.S. federal budget, but once the plants are removed, no money is left for environmental cleanup.[39]

Another hazardous waste problem with organized crime dimensions is the illicit traffic and smuggling of nuclear materials. Most notably, the lack of inexpensive, adequate, and safe disposal measures for radioactive waste is attracting criminal groups within Europe. Authorities there have investigated the illegal dumping of radioactive waste from Austria, France, Germany, and Eastern Europe into the Mediterranean and Adriatic seas by companies hired by Italian organized crime groups. In 1998, police investigated the 'Ndrangheta for dumping radioactive waste off Italy's southern coast.[40] Additional organized crime groups involved include Russian "mafiya" gangs, the Italian Mafia, and South African groups. But the typical nuclear materials smuggler may not be connected to organized crime at all—one expert holds that "the archetypal modern nuclear criminal is more likely to be the chief engineer or chief bookkeeper of a nuclear enterprise or the head of an import-export firm."[41] Russian customs officials have said that the diversion of nuclear materials has occurred through ostensibly legal channels.[42]

From 1991 to 1995, some 440 incidents were documented involving attempts to smuggle nuclear materials into Germany,[43] and between 1993 and 2007, the International Atomic Energy Agency (IAEA) documented 1,340 incidents of trafficking in radioactive and nuclear materials.[44] Typical smuggling routes are from the former Soviet republics, through Eastern Europe, into Germany, and on to clients in Libya, Iraq, Iran, Algeria, and Pakistan. Most of the reported thefts from the former Soviet republics have consisted of low-grade uranium, caesium-137, strontium-90, and cobalt-60—materials that cannot be used to build a bomb, but that are nonetheless environmentally hazardous.[45] In addition to the human and environmental costs associated with illicit disposal, non-weapons-grade radioactive substances can be used to poison aquifers or construct so-called "dirty bombs," with potentially disastrous results.[46]

RESPONDING TO THE PROBLEM

The United Nations Environmental Programme (UNEP) was the first entity to initiate action on the international waste trade. UNEP began drawing up guidelines in 1982, and the Cairo Guidelines on the Environmentally Sound Management of Hazardous Wastes was approved in 1987. The United States, the European Community (EC), and the Organization for Economic Cooperation and

Development (OECD) each established additional regulations on the cross-border transport of hazardous wastes in the mid-1980s. These various regulations were based on the principle of Prior Informed Consent (PIC), which stipulates that exporters must inform importers of the nature of the materials, and that the importers must voluntarily consent to the shipment.[47]

The most significant response to the increased transnational traffic in hazardous wastes is the Basel Convention (on the Control of Trans-boundary Movements of Hazardous Wastes and Their Disposal)—a comprehensive global environmental treaty that entered into force in 1992 and is comprised of 170 member countries, or Parties. The Convention was negotiated among 96 countries and 50 international organizations, including nongovernmental organizations (NGOs) such as Greenpeace. Waste producers and dealers naturally wanted to keep the trade legal but regulated, while many recipient countries and environmental groups wanted the international waste shipments banned altogether. UNEP sided with the developed world, and agreed that the trade should remain legal but regulated.[48]

Basel operates on a variation of the PIC principle: countries must be notified in advance of waste shipments, and importers must consent. Then, so long as the materials are to be disposed of in an "environmentally sound" manner, the shipment is legitimate. Parties are further required to enact domestic legislation to prevent and punish the illegal traffic in hazardous wastes, and are expected to minimize the quantity of hazardous wastes that cross borders. The Convention also requested that Parties restrict exports to those cases in which they can't dispose of the materials properly on their own, or if the waste is considered "raw material" for the importing nation. Trade with non-Parties is permitted if there is a bilateral or regional agreement in which wastes are disposed of safely. Wastes under Basel include used oil, biomedical waste, used lead-acid batteries, chemicals and pesticides, PCBs (compounds used in heat exchange fluids, electric transformers and capacitors, additives in paint, copy paper, and sealants and plastics), and other chemical wastes generated by industry and consumers. Radioactive waste is not included. The Convention also banned the shipment of hazardous wastes to Antarctica, and established a Secretariat to arrange periodic conferences with contracting Parties (the Secretariat also acts as a liaison center for information on waste management, and identifies illegal waste practices).[49]

In 1995, the Basel Convention was amended by the Basel Ban, which outlawed all forms of hazardous waste exports from the 29 wealthiest countries of the Organization of Economic Cooperation and Development (OECD) to all non-OECD members. However, the Basel Ban has yet to be ratified. In fact, the Basel Convention came into force without the participation of Africa, the EC (except France), the United States, and Japan. The United States has not yet ratified the Convention.[50]

Unfortunately, the Basel Convention has been easily circumvented due to a number of serious flaws. For one, key terms laying out rules and obligations of Parties are vaguely defined. "Environmentally sound management" is defined as "taking all practicable steps to ensure that hazardous wastes or other wastes are managed in a manner which will protect human health and the environment against the adverse effects which may result form such wastes."[51] Such language is wide open to subjective interpretation, and obviously very difficult to enforce. There is not even a clear definition of what constitutes hazardous waste, and toxic products such as banned pesticides are not included in the definition because they are not destined for disposal. Although the Convention lists properties of hazardous wastes, different definitions from country to country complicates matters, so that the boundary between legal and illegal, safe and unsafe disposal has not been clearly delineated. In addition, PIC procedures are weak and ineffective—letters of consent are not required to be sent to the Basel Secretariat for inspection, making it difficult if not impossible to verify that proper officials authorized shipments, or whether language was sufficiently clear for the importing country to provide informed consent. The Basel Secretariat has no power to monitor Parties or apply sanctions, the Convention lacks provisions for liability and compensation, and there is an absence of incentives to eliminate hazardous waste generation. Moreover, bilateral waste trade agreements are permitted between Parties and non-Parties.[52]

Perhaps the most significant circumvention of Basel rules has been the growth of waste exports to developing nations for the purpose of recycling—often a thin disguise for illegal dumping. Sometimes hazardous waste traffickers simply relabel their "products" as commodities bound for recovery efforts. Perhaps 90% of all waste exports to developing countries are designated for "recycling"; however, a large proportion of the materials designated for recycling are not recoverable, and must be disposed of in landfills or incinerated, or illegally dumped.[53]

Notorious examples of Basel circumvention are plentiful. Eastern and Central Europe, Latin America, and Asia are increasingly targeted by Western "recycling" export schemes. Waste exports to the African countries of Sierra Leone, Namibia, and Angola under "waste-to-energy" schemes are also common. Importing countries are offered aid packages for roads, health care, education, and incinerators, and agree to accept hazardous wastes to be used to run power plants. But the nations lack clear air regulations and the capacity to dispose of the wastes properly, so materials are incinerated and toxic chemicals simply released into the air. Another example of Basel circumvention involves British-owned Thor Chemicals, which has been importing mercury waste from Great Britain and the United States since the 1980s into South Africa. South Africa is not a party to Basel, and since mercury is considered "raw material" and not hazardous waste, the imports are technically legal. But mercury waste by-products from a Thor plant have leaked into the Umgeni River, which runs through the Zulu homeland.[54]

With the numerous limitations in the Basel Convention, additional attempts to effectively regulate the hazardous waste trade have been developed. After Basel, 69 African, Caribbean, and Pacific states (ACP) insisted that the EC ban exports of hazardous and radioactive wastes to states within the framework of the Lome IV Convention (an aid and trade convention between European states and the ACP that is periodically renegotiated). The Bamako Convention was signed by 12 African nations in 1991—the text was close to that of the Basel Convention, but was an important improvement because it banned hazardous waste imports, including radioactive wastes, into Africa. Bamako also bans all forms of ocean dumping of wastes, outlaws the importation of hazardous substances banned in the country of export, contains provisions on clean production methods within Africa, requires hazardous waste audits, and imposes rigid liability onto waste producers. Still, as impressive as it sounds, Bamako doesn't do much because it lacks the funding to monitor waste shipments—essentially, it is an unfunded mandate. In 1992, Central American states agreed on a ban similar to that of Bamako.[55]

The Stockholm Convention on Persistent Organic Pollutants (2001) entered into force in 2004. It identifies 12 pollutants slated for elimination. By August 2006, 127 countries had ratified the treaty (not the United States). The Stockholm Convention is especially significant because it seeks to ban POPs, chemicals that persist in the environment for decades, collect in the body fat of animals, and are transported up the food chain. Even low levels of exposure can cause developmental disorders in fetuses, damage to the immune and nervous systems, and a range of cancers. Most of the chemicals slated for elimination are pesticides, but dioxins, furans, and polychlorinated biphenyls (PCBs) are included as well. The United States has banned POPs and PCBs, but continues to export them—between 2001 and 2003, 28 million pounds of banned pesticides were exported from the United States.[56]

Another international regulatory effort is the Rotterdam Convention, signed by 110 countries. The treaty came into force in 2004, and provides controls on the international trade of various toxic chemicals. Countries importing listed toxic chemicals must be informed of bans and restrictions in the countries of export. The United States is not a signatory of the Rotterdam Convention.[57]

Another significant regulatory scheme involves the European Union Network for the Implementation and Enforcement of Environmental Law (IMPEL), an informal network of environmental authorities of EU member states, acceding and candidate states, and Norway. The European Commission is also a member. IMPEL lists three types of wastes to be regulated, each with its own set of controls (and opportunities for illegality):

- Green list—nonhazardous substances that can be traded more freely (not all low-hazard waste is green list; green list wastes include cadmium, lead, and some plastic defined as hazardous under Basel).
- Red list—hazardous wastes like PCBs that are subject to strict controls and the principle of Prior Informed Consent.
- Amber list—potentially hazardous but less risky than red-list materials—subject to "tacit" agreements, and may be shipped to some countries for recovery purposes only.[58]

In addition, IMPEL controls vary, and depend on the following:

- Purpose of the waste shipment—all trans-boundary movements of waste for disposal require notification, and many are absolutely prohibited (exports from IMPEL to non-OECD countries for disposal are prohibited).
- Type of waste—only green-list materials can be moved without notification controls (though not all of it).
- If the waste is destined for non-OECD countries, it is prohibited by the Basel Export Ban—even if for recovery, hazardous waste from OECD to non-OECD countries is banned by Basel.
- The countries of concern—green-list wastes from the European Union to OECD nations for recovery are subject to minimal controls; export to non-OECD countries are subject to different types of controls depending upon the nature of the waste, as well as the destination state and its particular requirements.[59]

Illegal waste shipments under the IMPEL regulations mostly involve waste wrongly classified as green list, or else green-list waste shipped to countries that lack the required controls. Clearly, the regulations are complicated, and some illegality is unintentional—merely a failure to understand and abide by the regulations. Conversely, for those with intent, circumvention is relatively easy. For example, since there is no legal trade, refrigerator equipment shipped from OECD countries to Africa is labeled as green-list materials destined for repair. However, many machines are actually beyond repair and simply intended for disposal—short of inspecting each unit for potential functionality, enforcement is nearly impossible.[60]

OZONE-DEPLETING SUBSTANCES
AND ELECTRONIC WASTE

While the Basel Convention and other efforts may have reduced the total export of toxic waste to developing countries to some extent, overall, the problem seems to be intractable. Moreover, new threats and challenges have emerged in recent years, most notably the illicit traffic in ozone-depleting substances (ODS)/chlorofluorocarbons and the rapidly growing amount of toxic electronic waste (e-waste) produced globally.

Ozone-depleting substances (chlorofluorocarbons, or CFC) are a common refrigerant used in cooling systems. Unfortunately, an international attempt to eliminate the use of CFC precipitated a significant black market. Under the Montreal Protocol's production and consumption control rules, the United States and other developed nations agreed to phase out CFC production and ban most imports. The problem is that developing nations have a longer time frame to phase out their production, and may still legally produce CFC for use in older machines that are not adaptable to other coolants. Since there is a strong demand in the developed world for replacement CFC in existing machines, a strong economic incentive for an illicit international traffic was generated. The black market is fueled by the fact that illegal imports are far cheaper than legally recycled CFCs or those materials obtained from limited existing stocks.[61]

As of 2003, about 20,000 to 30,000 metric tons of CFC was smuggled annually, half of which entered the United States (CFCs may be the second-largest illegal import into the United States after drugs). Other recipient states include nations throughout the European Union, as well as Canada and Taiwan.[62] The traffic generates some $1–2 billion a year, and involves, in part, Chinese, Latin American, and Russian crime groups. The primary methods used by smugglers are false labeling, the production of counterfeit paperwork, and the use of bogus export corporations.[63]

If the Montreal Protocol has been a failure, there have nevertheless been some law enforcement successes in combating the traffic in CFC/ ODS. For example, in a 2003 sting, operatives posing as clients found a Singapore company willing to source and ship CFC-12 to South Africa —the company said it could repackage the goods, supply false labels and documents, and ship the product to neighboring countries.[64] In the United States, the loss of tax revenue generated swift action: "Operation Cool Breeze" consisted of a joint operation by the Environmental Protection Agency, the Internal Revenue Service, the Department of Justice, and the Commerce Department. In all, 500 tons of CFCs valued at $40 million were seized (and handed over to the Department of Defense, which uses it to recharge older equipment). Cool Breeze also precipitated the first international extradition of an individual for committing an environmental crime. Unfortunately, a notable success like Cool Breeze is counterbalanced by a lax enforcement effort in the European Union due to the lack of a tax recovery incentive.[65]

One of the more severe hazardous waste issues is the increasing deluge of global e-waste, including broken or obsolete computer components, cell phones, cathode-ray tubes, DVD players, VCRs, copiers, fax machines, stereos, and video gaming systems. Each year, almost 7 million tons of high-tech electronics become obsolete in the United States alone, while the EPA says more than 4.6 million tons of e-waste was discarded in landfills in 2000.[66] In fact, most e-waste ends up in landfills, incinerators, and ill-equipped recycling facilities in developing nations. In Asia, Africa, and Latin America, workers disassemble components for resale or for use in new manufacturing processes— but then the hazardous components are simply dumped.[67] Inspections of 18 European seaports in 2005 found that 47% of waste (including e-waste) destined for export was illegal, and in 2003 in the UK, 23,000 metric tons of undeclared gray-market e-waste was transported to the Far East, India, Africa, and China (see Table 2.1).[68]

The amount of e-waste produced each year is prodigious—the average lifespan of computers dropped from six years in 1997 to two years in 2005, and cell phones are often discarded in less than two years in developed countries. Between 1997 and 2004, 315 million computers containing 1.2 billion pounds of lead became obsolete. Estimates vary, but perhaps 20–50 million tons of e-waste is generated annually, and it comprises about 5% of all municipal solid waste worldwide (about the same amount as plastic packaging).[69] Improper disposal poisons land and water with a number of toxins, including lead, arsenic, cadmium, antimony trioxide, polybrominated flame retardants, selenium, chromium, cobalt, and mercury. When not dumped, smelting and burning of these wastes simply releases the toxins into the air.[70]

Table 2.1. California and EPA Data: 2007 Estimated E-waste Exports (Kilos) by Designated Country

California Department of Toxic Substances Control	Kilos (in thousands)	EPA Notifications of Broken CRT Export	Kilos (in thousands)
Malaysia	3,583	Malaysia	50,699
Canada	Not reported	Canada	11,175–11,689
Brazil	1,633	Brazil	3,428–1,099
South Korea	1,588	South Korea	7,103
China	1,043	China	NR
Mexico	816	Mexico	NR
Vietnam	318	Vietnam	NR
India	91	India	NR

Source: *Electronic Waste and Organized Crime: Assessing the Links*, Phase II Report for the INTERPOL Pollution Crime Working Group, May 2009, 11, http://www.interpol.int/Public/ICPO/FactSheets/WasteReport.pdf (accessed September 19, 2010).

Exposure to the toxins in e-waste can be devastating to humans. Lead causes damage to the brain and nervous system, blood disorders, and kidney damage, and leads to developmental damage in fetuses. Cadmium, a carcinogen, can also cause kidney damage and damage to bone structure. Beryllium causes lung cancer and chronic beryllicosis; mercury causes brain and kidney damage; and trichloroethylene and trichloroethane are toxic to nervous, respiratory, endocrine, and reproductive systems as well as kidney and liver functions. About 20% of the average computer is comprised of polyvinyl chloride (PVC)—incineration creates dioxins and furans that cause cancer, immune suppression, liver damage, hormonal disruptions, and behavioral changes.[71] Brominated flame retardants used in circuit boards and plastic casing interfere with hormone systems and the thyroid, and exposure to such chemicals while in the womb has been linked to behavioral problems and impaired learning and memory loss (about 1,000 tons of such materials were used to manufacture 674 million cell phones in 2004; see Table 2.2).[72]

While disposal in landfills and incinerators causes pollution, improper recycling of secondhand electronics poses risks as well. In Asia, recycling is often done in scrapyards by children, and plastic

Table 2.2. Estimated Amounts of E-waste Generated

E-waste Source	Time Period	Amount	Type of E-waste Included	Source of Estimate
International	2007	50 million tons	Personal computers; e-waste	INTERPOL Pollution Crime Working Group Phase II Report (citing UNEP)
United States	2006	21 million tons	E-waste	Centillion Environment and Recycling (citing EPA)
United States	2002	12.5 million tons	E-waste	http://www.ban.org (citing Carnegie Mellon University)
United States	2005	2 million tons	E-waste	EPA 2007
United States	2005	175,000 tons	CRTs collected for recycling	EPA 2007
United States	2008	300,000 to 400,000 tons collected annually	Electronics	http://www.abcmoney.co.uk
Canada	2000 and 2003	140,000 tons	Computer equipment, phones, audio-visual equipment, small household appliances	Environment Canada at http://www.ec.gc.ca/envirozine/english/issues/33/
Nigeria—destination	2006	6,000 40-foot containers annually	Used electronics; 75% estimated as unsalvageable	http://www.ban.org

Source: *Electronic Waste and Organized Crime: Assessing the Links*, Phase II Report for the INTER-POL Pollution Crime Working Group, May 2009, 13, http://www.interpol.int/Public/ICPO/FactSheets/WasteReport.pdf (accessed September 19, 2010).

e-waste is incinerated with no controls—furans and dioxins are released into the atmosphere.[73] In one Chinese town where children are used in primitive e-waste recycling efforts, blood lead levels are significantly higher than in neighboring towns. In southern Taiwan, the Erren River is home to illegal and/or improper e-waste recycling facilities—fish there die within two minutes of exposure to river water, and humans experience unnaturally high cancer rates.[74]

The global problem of electronic waste is exacerbated by the fact that the profits from unsafe and illegal disposal are tremendous. First, zero waste recycling and zero waste disposal of electronic equipment is expensive: it costs about $18 to remove the lead from just one computer monitor or television screen. However, electronic equipment also contains precious metals, including nickel, copper, iron, silicon, and gold (cell phones are 19% copper and 8% iron), which may be extracted as well. But this too is labor intensive and expensive, so local authorities anxious to meet recycling targets and manufacturers obligated under producer-responsibility regulations are happy to allow e-waste handlers to assume the responsibility of collection, transport, and disposal. Unscrupulous outfits often charge very little, even nothing, for the e-waste—reputable U.S. electronics recyclers who charge a fee for disposal state that entities that offer free disposal or even pay for electronic waste must be disposing of it improperly because they could not stay in business otherwise.[75] Once collected, legitimate and illegal operators usually transport the materials to developing nations, who welcome the "recycling" revenue. The profit comes not only from the extraction of the precious metals, but also from the large price differentials between developed and Third World countries. For example, glass-to-glass recycling of computer monitors costs 50 cents per pound in the United States, but only five cents in China—Third World recycling companies pay their workers (often children) low wages, are typically unconcerned with safety or health measures, and are not burdened by stringent environmental rules. Again, after recycling or refurbishment, the poisonous e-waste leftovers are routinely dumped in rivers and placed in unlined landfills, or are improperly incinerated.[76]

Illicit traffickers typically mislabel containers and mix electronic components with legitimate consignments. In the United Kingdom "e-waste tourists" visit the country to purchase electronic waste so that they can extract precious metals, but then dump the leftovers because proper disposal would eliminate their profits. Researchers have observed numerous methods of illegal e-waste disposal in the United Kingdom:

- Direct illegal disposal, called "fly tipping"
- The use of unlicensed waste management sites
- The use of unlicensed carriers, brokers, or waste tourists (who travel to exporting countries and buy up e-waste for export or resale, and then help dispose of it illegally)

- Shipping infringements, such as providing false shipment details to customs officials, using false labels (calling electronic waste "personal effects" or "used goods"), mixing waste with end-of-life vehicles, and shipping waste to a fictitious address
- Abuse of recycling initiatives (using "recycling" as cover for illicit activities), including unregulated recycling and other illicit disposal activities; this category includes smelting to extract precious metals, fly tipping, and unregulated dismantling[77]

Extracting precious metals under the guise of recycling is especially common, and then the toxins are dumped. Of course, large amounts of e-waste is unsalvageable, and even when equipment is saved, the remainder is typically disposed of improperly. Enforcement in this area is especially difficult because one would have to prove that items labeled for "recovery" were actually waste (each electronic component would have to be tested for functionality).[78]

In the United States, there are at least 2,000 firms involved in collecting e-waste. Almost all of this is destined for export, as the United States lacks appropriate recycling facilities for some materials, and doesn't have any smelters for copper and precious-metal recovery or CRT glass furnaces. (In fact, there are only five copper/precious metal smelters in the world equipped to minimize the release of dioxins—they are in Canada, Sweden, Germany, Belgium, and Japan.) Some exporters are small operators, while others are part of large U.S.-based corporate structures. In some cases, a company will use a subsidiary to export waste to a foreign subsidiary (the company exports waste to itself). The majority of exporters ship to Canada, but by volume, most U.S.-produced e-waste goes to Malaysia, with smaller but still significant shipments to Brazil, South Korea, China, and Mexico. Internationally, e-waste exports usually pass through one or more ports, with Rotterdam being one of the most important. Typically, criminals are based outside of developed countries and visit to secure the e-waste for export, or else use smaller operators in the country of origin to organize collection and shipment.[79]

The regulation of e-waste is complex and difficult to enforce. Under the Basel Convention, e-waste is hazardous and therefore falls under the Basel Ban. The OECD uses it own control system based on tacit consent as opposed to prior informed consent. In addition, the OECD definition of hazardous waste is based on risk, while the Basel definition is based on the presence of toxins. As of 2007, the OECD control system classified CRTs as hazardous waste subject to controls and consent, but circuit boards were still defined as commodities.[80] International law *generally* prohibits the export of electronic waste for *disposal* purposes. While the transport of hazardous waste from OECD to non-OECD countries is not permissible, the export of nonhazardous wastes for recovery of raw materials or for the second-hand market is allowed (which means television and computer monitors may be exported, but not for salvage or parts).

U.S. regulations pertaining to e-waste are far less stringent than international laws. For example, many items listed as hazardous under Basel are classified as nonhazardous or non-waste under U.S. law. (As mentioned earlier, the United States has not ratified the Basel Convention.) While the United States cannot export waste to non-OECD countries without a bilateral agreement, it has in fact implemented bilateral waste trade agreements with Mexico and Canada.[81] The principal U.S. legislation governing e-waste is the Resource Conservation and Recovery Act (RCRA), which only regulates materials that are classified first as waste, and then as *hazardous* waste. The RCRA does not address e-waste as a unique form of hazardous waste—under existing law, materials that are commonly understood to constitute e-waste is largely unregulated because it is classified as nonhazardous, or as non-waste. So in the United States, much e-waste is either exempt or excluded from environmental regulations. Waste excluded from regulation includes electronic equipment designated for reuse and materials that can be recycled into new products (for example, processed scrap metal, shredded circuit boards, CRT glass, and intact CRTs). However, most e-waste cannot legally be placed in U.S. landfills.[82]

China banned the import of e-waste in 2000, but the law is routinely violated. In Europe and Japan, costs for e-waste disposal have been shifted from taxpayers and the government to manufacturers. In 2002, the EU implemented two directives: the Restriction of Hazardous Substances (RoHS) and Waste from Electrical and Electronic Equipment (WEEE). The RoHS mandated that electronics manufacturers stop using toxic chemicals and heavy metals in their products—a ban that includes cadmium, mercury, lead, hexavalent chromium, and two types of brominated flame retardants. The WEEE directive orders that producers must take back their electronic products when they are discarded. Since 2001, Japanese manufacturers must take back refrigerators, washing machines, air conditioners, televisions, and, more recently, personal computers.[83]

ASSESSING VULNERABILITY AND RESPONDING STRATEGICALLY

If the problem of illicit waste trafficking is not intractable, solutions nonetheless remain elusive. Even when international regulations are implemented, new markets soon appear. In the 1980s, traffickers responded to tightened European regulations by shifting destination routes from Central Europe to Africa and Asia, and when the Bamako Convention banned the import of hazardous waste into various African, Caribbean, and Pacific nations, traffickers adapted once more by switching routes to China and Eastern Europe.[84] The ad hoc and fluid nature of many criminal networks also impedes law enforcement—if authorities focus on 40-foot shipping containers, traffickers switch to smaller conveyances, even open-sided trucks.[85] Additional factors that contribute to the expansion of illicit waste trafficking are large-scale geopolitical changes like the transformation of the former Soviet bloc countries, as well as the evolution of global trade in the direction of liberalization, deregulation, and the proliferation of transnational corporations. New multilateral environmental agreements provide new opportunities for evasion (many countries simply lack the capability to enforce agreements),[86] and of course, the problem is further exacerbated by weak or nonexistent environmental laws in developing nations, and corruption at high levels of government.[87]

Another factor that increases the likelihood of criminality in the garbage and hazardous waste industry is that waste is a product with a low level of integrity, meaning that the physical nature of the product is such that it is easily altered for the purpose of deception.[88] The huge legal trade in recyclable materials (such as scrap metal) facilitates the commingling of hazardous wastes with legal materials, and provides convenient cover for traffickers who look to transport hazardous waste as recoverable secondhand goods. Disguising hazardous waste as a legitimate product is common. For example, in 1992, Bangladesh received 1,000 tons of copper smelter furnace dust with high levels of lead and cadmium. The waste was mixed with fertilizer by U.S. firms and individuals, and then was sold to Bangladesh with the aid of the Asian Development Bank. Before the scam was uncovered, the toxic waste had already been spread on farmland.[89]

Additional factors that facilitate the trade in hazardous waste are inelastic demand for the product or service and the emergence of waste brokers. Inelastic demand in the waste disposal industry ensures that increasing prices for disposal do not reduce demand for the service—this attracts unsavory businesspeople, including organized crime groups. Waste brokers emerged as a mechanism to help industries comply with more stringent environmental regulations. However, because the brokers do not become legal owners of garbage and hazardous waste, they more easily escape legislative and regulatory control. In such ways, the waste becomes "orphaned," and no one assumes full responsibility.[90]

The high costs for disposal and regulatory failure are but one part of the risk equation, as there are enforcement-level issues such as insufficient resources, a desire to avoid political and economic disruptions, and outright corruption.[91] An additional difficulty in enforcement and prosecution is a bifurcation between administrative law and penal law, with different spheres of responsibility and operation. Law enforcement controls focus on the transport of waste, and administrative controls deal with licensing and site inspection—this split hampers investigations and prosecutions. As noted, environmental laws are very complex, and criminal justice officials have difficulty negotiating the dynamics of overlapping national and international regulations. Moreover, both penal and administrative controls are oriented nationally, but the waste trade has become increasingly transnational in scope.[92]

Another issue that contributes to vulnerability in the waste disposal industry is merger activity resulting in market concentration (a small number of companies account for market share). The remaining smaller firms are highly vulnerable to infiltration by organized crime, largely due to competitive pressures from the larger companies.[93] Barriers to entry are low but rising—new waste companies must have contacts, technical competency, and the ability to meet environmental regulations as well as an administrative capacity capable of meeting licensing requirements—a good thing, as this facilitates transparency. However, some new and smaller market players are less transparent, and may have to rely on largely unregulated waste brokers to fulfill business needs—bringing smaller and newer firms together with waste brokers deepens vulnerability to illicit actions such as unsafe disposal.[94]

One final, very real danger is that industries that produce a large amount of waste will continue to move their operations to developing nations with cheap labor and loose regulations. Some waste disposal services are entirely exported (or outsourced) to Third World nations, a prime example being the practice of shipbreaking (tearing down old ships uncovers a huge amount of toxic waste, including asbestos).[95]

Since 1996, G8 summits have called for coordinated actions to combat international environmental crime, and Interpol, the World Customs Organization, and UNEP have all worked on the issue. Networks of environmental enforcement have also been established, including the International Network of Environmental Compliance and Enforcement and the European Network on the Implementation and Enforcement of Environmental Law. The Lusaka Agreement (1994) between six African countries was implemented to enhance the cross-border enforcement of wildlife law, and may serve as a good model for waste traffic regulators.[96]

Still, international measures, most notably the Basel Convention, have been largely ineffective in combating illicit waste disposal and trafficking. High costs associated with disposal in relatively wealthy nations and lax regulations (and cheaper disposal) in developing nations drive the black market in garbage and hazardous wastes. Inconsistent regulations, public corruption, and a dearth of political will and international cooperation exacerbate the problem and render even well-constructed regulations impotent. If any progress is to be made on the political and criminal justice front, international environmental regulations must be consistent and binding, and considerably greater resources should be devoted to enforcement.

Law enforcers and environmental regulators must adopt a more strategic approach, and recognize that the illegal movement of wastes should not be viewed as a singular problem, but as an issue with multiple dimensions. For example, specific kinds of waste emanating from the European Union have different destinations: plastic waste tends to be exported to Asia, refrigerators and CFC tend to go to Western Africa, end-of-life vehicles go to Africa and Eastern Europe, and electronic and cable waste is shipped to West and East Asia. Likewise, the *modus operandi* also differs depending upon the type of waste, the origin of the waste, and the destination of the hazardous material. Data from England and Wales shows some overlap between licit and illegal transit routes for some kinds of waste like plastics, but there is no overlap in relation to other wastes like refrigerators. Illicit exporters ship different types of waste using different routes to different countries, and different groups are involved in the various waste exports—in the case of WEEE exports to Africa, foreign nationals from the destination state make the export from the origination state; but for other wastes, nationals within the exporting country are responsible.[97] In sum, criminals respond and adapt to differential opportunities based on factors such as regulation and pricing, so law enforcers must be prepared to respond in kind.

While addressing illicit waste disposal and trafficking on a transnational scale presents numerous difficulties, strategies at the national and local levels may be relatively simple to implement. Some answers may very well lie outside the realm of criminal justice, as national, regional, and local governments have the power to significantly modify the private behaviors of businesses and individuals minus the threat of fines or imprisonment. Incentives such as public subsidies could be used as the "carrot" to encourage legal disposal of garbage, while disincentives such as taxation could serve as an effective "stick." In the Philippines, high levels of submission to environmental laws have been observed when municipal mayors and the Municipal Environment and Natural Resources Officer (MENRO) work closely together to monitor compliance at the neighborhood, and even the household level.[98] Incentivizing countless individuals to properly dispose of their garbage could, *to some degree*, obviate the need for ineffective transnational bodies to police organized waste traffickers and corrupt public officials.

Finally, technological innovations may fill the regulatory void by indirectly reducing the profitability of criminals who traffic in garbage and hazardous wastes. Clearly, reducing costs for local treatment and disposal of nonrecyclable waste and increasing prices paid for recyclable wastes would reduce the illegal waste trade. Demand reduction strategies and methods that limit the volume of waste and reduce the costs of responsible disposal will necessarily decrease *opportunities* for illicit traffickers.

NOTES

1. Jennifer Clapp, "The Illicit Trade in Hazardous Wastes and CFCs: International Responses to Environmental Bads," *Trends in Organized Crime* 3, no. 2 (1997): 14–18; Christoph Hilz, *The International Toxic Waste Trade*, (New York: Van Nostrand Reinhold, 1992); Bill D. Moyers, *The Global Dumping Ground*, (Santa Ana, CA: Seven Locks Press, 1990); "Transboundary Movements of Hazardous Wastes: the Case of Toxic Waste Dumping in Africa," *International Journal of Health Services* 21, no. 4 (1991): 759–77; Laura A. Strohm, "The Environmental Politics of the International Waste Trade," *Journal of Environment and Development* 2, no. 2 (Summer 1993): 129–53; Donald J. Rebovich, *Dangerous Ground: The World of Hazardous Waste Crime* (New Brunswick, NJ: Transaction Publishers, 1992).

2. Monica Massari and Paola Monzini, "Dirty Business in Italy: A Case Study of Trafficking in Hazardous Waste," *Global Crime* 6, no. 3–4 (2004): 285–304.

3. Ibid.

4. Clapp, "Illicit Trade in Hazardous Wastes and CFCs."

5. "International Crime Threat Assessment," *Trends in Organized Crime* 5, no. 4 (2000): 32–144; Massari and Monzini, "Dirty Business in Italy."

6. Sarah Czarnomski, Barry Webb, and Alan Holmes, "IMPEL-TFS Threat Assessment Project: The Illegal Shipment of Waste Among IMPEL Member States," Jill Dando Institute of Crime Science, University College of London, 2005, http://impel.eu/wp-content/uploads/2010/02/2006-x-Threat-Assessment-Final-Report.pdf.

7. Jennifer Clapp, "Africa, NGOs, and the International Toxic Waste Trade," *Journal of Environment and Development* 3, no. 2 (1994): 17–46.

8. Ibid.

9. Ibid.

10. Ibid.

11. Clapp, "Illicit Trade in Hazardous Wastes and CFCs."

12. "The Digital Dump: Summary of Findings," BAN Report, October 24, 2005, http://www.ban.org/BANreports/10-24-05/documents/ SummaryofFindings.htm (accessed September 19, 2010).

13. Clapp, "Africa, NGOs, and the International Toxic Waste Trade."

14. "China Set to Curb Foreign Waste Imports," China Daily, January 24, 2007, http://www.chinadaily.com.cn/china/2007-01/24/content_791722 .htm (accessed September 19, 2010).

15. "Smugglers Find Treasure in Trash," People's Daily Online, December 29, 2007, http://english.peopledaily.com.cn/90001/90776/6329761.html (accessed September 19, 2010).

16. "Guangzhou Customs Intercepts 237 Tons of Smuggled Trash," China.org.cn, 2007, http://japanese.china.org.cn/english/SO-e/27771.htm (accessed September 19, 2010).

17. "Smugglers Find Treasure."

18. Massari and Monzini, "Dirty Business in Italy," 289.

19. Massari and Monzini, "Dirty Business in Italy."

20. Ibid.

21. Robin Pomeroy, "Naples Garbage Is Mafia Gold," Reuters, January 9, 2008, http://www.reuters.com/article/idUSL083057720080109 (accessed September 19, 2010); Roberto Saviano, *Gomorrah: A Personal Journey into the Violent International Empire of Naples' Organized Crime System* (New York: Farrar, Straus, and Giroux, 2007).

22. "International Crime Threat Assessment."

23. "Nuclear News: EPA to Rebuild Uranium-Contaminated Navajo Homes," June 15, 2009, http://weblog.greenpeace.org/nuclear-reaction/ 2009/06/nuclear_news_epa_to_rebuild_ur.html (accessed September 19, 2010).

24. "Mafia Sank Boat with Radioactive Waste: Official," Nuclear Power Daily, September 14, 2009, http://www.nuclearpowerdaily.com/reports/ Mafia_sank_boat_with_radioactive_waste_official_999.html (accessed September 19, 2010).

25. "Mob Expert Says Naples Garbage Fix Only Temporary," FoxNews .com, January 9, 2008, http://www.foxnews.com/story/0,2933,321410,00 .html (accessed September 19, 2010).

26. Monica Massari, "Ecomafias and Waste Entrepreneurs in the Italian Market," paper presented at the 6th Cross-border Crime Colloquium, Berlin, Germany, 2004.

27. Alan A. Block, *The Business of Crime* (Boulder, CO: Westview, 1991); Alan A. Block and Frank R. Scarpitti, *Poisoning for Profit: The Mafia and Toxic Waste Disposal in America* (New York: William Morrow, 1985); U.S. House of Representatives, 97th Congress, 1st sess., *Organized Crime Links to the Waste Disposal Industry* (Washington, DC: U.S. Government Printing Office, 1981); Peter Reuter, Jonathan Rubenstein, and Simon Wynn, *Racketeering in Legitimate Industries: Two Case Studies*; (Washington, DC: National Institute of Justice, 1983); U.S. Senate, Permanent Subcommittee on Investigations of the Committee on Governmental Affairs, 98th Congress, 1st sess.,

Profile of Organized Crime: Mid-Atlantic Region (Washington, DC: U.S. Government Printing Office, 1983).

28. U.S. House of Representatives, *Organized Crime Links to the Waste Disposal Industry*; James B. Jacobs, Christopher Panarella, and Jay Worthington, *Busting the Mob: U.S. v. Cosa Nostra* (New York: New York University Press, 1994); Comptroller General's Report to the Subcommittee on Investigations and Oversight, Committee on Public Works and Transportation, U.S. House of Representatives, *Illegal Disposal of Hazardous Waste: Difficult to Detect or Deter* (Washington, DC: U.S. General Accounting Office, 1985).

29. Hearings held by the Subcommittee on Oversight and Investigations, Committee on Interstate and Foreign Commerce, U.S. House of Representatives, *Organized Crime and Hazardous Waste Disposal* (Washington, DC: U.S. Government Printing Office, 1980).

30. Andrew Szasz, "Corporations, Organized Crime, and the Disposal of Hazardous Waste: An Examination of the Making of a Criminogenic Regulatory Structure," *Criminology* 24, no. 1 (1986): 1–27.

31. Ibid.

32. Nicholas Dorn, Stijn Van Daele, and Tom Vander Beken, "Reducing Vulnerabilities to Crime of the European Waste Management Industry: The Research Base and the Prospects for Policy," *European Journal of Criminology, Criminal Law, and Criminal Justice* 15, no. 1 (2007): 23–36.

33. Szasz, "Corporations, Organized Crime, and the Disposal of Hazardous Waste."

34. Massari and Monzini, "Dirty Business in Italy," 298.

35. *Electronic Waste and Organized Crime: Assessing the Links*, Phase II Report for the INTERPOL Pollution Crime Working Group, May 2009, http://www.interpol.int/Public/ICPO/FactSheets/WasteReport.pdf (accessed September 19, 2010).

36. Gruppo Abele-Nomos, *The Illegal Trafficking in Hazardous Waste in Italy and Spain: Final Report*, 2003, http://www.organized-crime.de/revgru03.htm (accessed November 1, 2010).

37. United Nations Drug Control Programme, "Environmental Effects of Illicit Crop Cultivation," *Trends in Organized Crime* 3, no. 2 (1997): 11–14.

38. Ibid.

39. "Pot Farms Wreak Havoc in Sequoia National Park," 2005, http://seattletimes.nwsource.com/html/nationworld/2002445217_marijuana19.html (accessed September 19, 2010).

40. "International Crime Threat Assessment."

41. Rensselaer Lee, "Recent Trends in Nuclear Smuggling," *Transnational Organized Crime* 2, no. 4 (1996): 114.

42. Lee, "Recent Trends in Nuclear Smuggling."

43. Joseph Albini, R. E. Rogers, Victor Shabalin, Valery Kutushev, Vladimir Moiseev, and Julie Anderson, "Russian Organized Crime: Its History, Structure, and Function," *Journal of Contemporary Criminal Justice* 11, no. 4 (1995): 213–43.

44. "Illicit Trafficking and Other Unauthorized Activities Involving Nuclear and Radioactive Materials," 2007, http://www.iaea.org/NewsCenter/Features/RadSources/PDF/fact_figures2007.pdf (accessed November 1, 2010).

45. Brian Freemantle, *The Octopus: Europe in the Grip of Organized Crime* (London: Orion, 1995).

46. Lee, "Recent Trends in Nuclear Smuggling."

47. Clapp, "Africa, NGOs, and the International Toxic Waste Trade."

48. Ibid.

49. Secretariat of the Basel Convention, "The Basel Convention at a Glance," http://www.basel.int/convention/bc_glance.pdf (accessed September 19, 2010).

50. *Electronic Waste and Organized Crime: Assessing the Links.*

51. Clapp, "Africa, NGOs, and the International Toxic Waste Trade," 31.

52. Clapp, "Africa, NGOs, and the International Toxic Waste Trade."

53. Ibid.

54. Ibid.

55. Ibid.

56. Ibid.

57. Kristin S. Schafer, "One More Failed U.S. Environmental Policy," Foreign Policy in Focus, August 31, 2006, http://www.fpif.org/reports/one_more_failed_us_environmental_policy (accessed September 19, 2010).

58. Czarnomski et al., "IMPEL-TFS Threat Assessment Project."

59. Ibid.

60. Ibid.

61. Clapp, "Illicit Trade in Hazardous Wastes and CFCs"; Gavin Hayman and Duncan. Brack, *International Environmental Crime: The Nature and Control of Black Markets* (London: The Royal Institute of International Affairs, 2002).

62. Ibid.

63. "International Crime Threat Assessment."

64. Charles W. Schmidt, "Environmental Crimes: Profiting at the Earth's Expense," *Environmental Health Perspectives* 112, no. 2 (February 2004): A96–A103.

65. Clapp, "Illicit Trade in Hazardous Wastes and CFCs."

66. "Toxic Tech: Pulling the Plug on Dirty Electronics," Greenpeace International, May 2005, http://www.greenpeace.org/raw/content/international/press/reports/toxic-tech-puling-the-plug-o.pdf (accessed September 19, 2010); Elizabeth Grossman, *High Tech Trash: Digital Devices, Hidden Toxics and Human Health* (Washington: Island Press/Shearwater Books, 2006).

67. David Naguib Pellow, *Resisting Global Toxics: Transnational Movements for Environmental Justice* (Cambridge, MA: MIT Press, 2007).

68. *Electronic Waste and Organized Crime: Assessing the Links.*

69. "Toxic Tech: Pulling the Plug on Dirty Electronics."

70. *Electronic Waste and Organized Crime: Assessing the Links*; "Mobile Toxic Waste: Recent Findings on the Toxicity of End-of-Life Cell Phones" (Basel Action Network, 2004).

71. "Toxic Tech: Pulling the Plug on Dirty Electronics"; *Electronic Waste and Organized Crime: Assessing the Links.*

72. "Toxic Tech: Pulling the Plug on Dirty Electronics."

73. Ibid.

74. *Electronic Waste and Organized Crime: Assessing the Links.*

75. Ibid.

76. Ibid.

77. Ibid.

78. Ibid.

79. Ibid.

80. Robert Tonetti, "Export of Used and Scrap Electronics: What You Need to Know," http://www.epa.gov/osw/conserve/materials/ecycling/docs/exports.pdf (accessed September 19, 2010).

81. *Electronic Waste and Organized Crime: Assessing the Links.*

82. Ibid.

83. Ibid.

84. Massari and Monzini, "Dirty Business in Italy."

85. Czarnomski et al., "IMPEL-TFS Threat Assessment Project."

86. Dorn, et al., "Reducing Vulnerabilities to Crime"; Duncan Brack, "The Growth and Control of International Environmental Crime," *Environmental Health Perspectives* 112, no. 2 (2004): 80–81.

87. Clapp, "Illicit Trade in Hazardous Wastes and CFCs."

88. Dorn et al., "Reducing Vulnerabilities to Crime."

89. Clapp, "Illicit Trade in Hazardous Wastes and CFCs."

90. Dorn et al., "Reducing Vulnerabilities to Crime."

91. Brack, "Growth and Control of International Environmental Crime"; Timothy S. Carter, "The Failure of Environmental Regulation in New York: The Role of Cooptation, Corruption and Cooperative Enforcement Approach," *Crime, Law and Social Change* 26, no. 1 (1996): 27–52.

92. Dorn et al., "Reducing Vulnerabilities to Crime."

93. Andrew Cooke and Wendy Chapple, "Merger Activity in the Waste Disposal Industry: The Impact and the Implications for the Environmental Protection Act," *Applied Economics* 32, no. 6 (2000): 749–55.

94. Dorn et al., "Reducing Vulnerabilities to Crime."

95. K. Paul, "Exporting Responsibility: Shipbreaking in South Asia: International Trade in Hazardous Waste," *Environmental Policy and Law* 34, no. 2(2004): 73–78.

96. Brack, "Growth and Control of International Environmental Crime."

97. Czarnomski et al., "IMPEL-TFS Threat Assessment Project."

98. Ma. Eugenia C. Bennagen, Georgina Nepomuceno and Ramil Covar, "Solid Waste Segregation and Recycling in Metro Manila: Household Attitudes and Behavior" (Singapore: Economy and Environmental Program for Southeast Asia, 2002), http://idl-bnc.idrc.ca/dspace/bitstream/10625/32393/3/118095.pdf (accessed November 1, 2010).

CHAPTER 3

The Illicit Traffic in Wildlife

The trade in wild animals and plants is global in scope, driven by prodigious demand for live specimens, animal parts, and derivative products. Much of this trade is illicit, and includes the traffic in exotic plants and endangered animal species. The consequences of the illicit trade are significant, with potentially severe environmental and human impacts, including the extinction of species, a reduction in biodiversity, and the spread of disease. Over the last 40 years in Vietnam, 12 large animal species have become extinct or virtually extinct, mainly due to the illicit wildlife trade.[1] In 1996, between $10 billion and $15 billion in exotic species were traded illegally,[2] while a more recent estimate from the World Wildlife Fund put the figure of illicit profits from animal and wildlife trafficking at $15–25 billion.[3] The trade is so lucrative that, as an illicit industry, it trails only drug and arms trafficking in significance.[4]

Approximately 25,000–30,000 primates, 2–5 million birds, 10 million reptile skins, 7–8 million cacti, 500 million tropical fish, and unknown numbers of plants and invertebrates (including leeches, spiders, insects, and corals) are removed from the wild annually for use as pets, entertainment, food, clothing, traditional medicines, or for specialist collections—an unknown but significant portion of specimens are trafficked in contravention of national and international laws.[5] In the United Kingdom alone, over one million plants, live and dead animals, animal parts, and medicines produced from endangered species were seized during a recent 12-month period.[6] The United States is also a major consumer of illegally trafficked animals, and China is the largest market in the world for tiger bone, rhino horn, ivory, and sea horse. The illicit wildlife trade in India is estimated at $1 billion,[7] while the European Union is the world's leading destination for reptile skins, parrots, and boa and python snakes. There are also indications the problem is increasing—in Mexico in 2007, 206,828 animals were confiscated, a number 110 times greater than in 2001. Law enforcement efforts have been only marginally successful, while the regulation of the trade naturally fosters the further development of black markets—in 1999, a moratorium on ivory harvesting for the Japanese market led to an immediate and massive increase on the poaching of elephants in Africa.[8]

Snakes and tortoises are especially popular commodities because of their ability to survive long trips, but the trade includes all manner of species living and dead: the horns of endangered black rhinos, Brazilian monkeys, Australian birds, elephant tusks, the bones of tigers, and exotic skins used for designer clothing.[9] By any standard, the trade is characterized by cruelty: during transport, toucans have their beaks taped shut, parrots are stuffed into stockings, and birds are drugged and have their eyes perforated so they won't sing in reaction to the light. Perhaps the majority of animals die during the smuggling process, a fact of the illicit trade that precipitates even greater harvesting of already stressed or endangered species. According to the National Network to Fight Wild Animal Trafficking, perhaps 38 million animals are caught annually in Brazil, but 90% die in the process of capture and transport.[10] Moreover, the problem of overharvesting animals and plants has implications beyond the extinction of certain species, as the decimation of habitats can lead to the collapse of entire ecosystems.[11] In addition to the severe environmental consequences of species destruction and the reduction in biodiversity, the unlawful traffic in wildlife has a significant potential to spread disease across species lines (for example, SARS and avian influenza).[12]

High profits and minimal risks produce a highly attractive environment for illegal entrepreneurs—a single Komodo Dragon is worth $30,000, a bluefin tuna, up to $100,000.[13] While some offenders are linked to or are themselves legitimate traders, established organized crime groups are increasingly involved in the illicit traffic of plants and animals.[14] In the United Kingdom, 50% of those prosecuted for wildlife crimes have previous convictions for offenses such as drugs and guns. In Brazil, a parliamentary commission has documented the link between animal trafficking and the traffic in drugs and precious stones, while Russian organized crime groups are believed to control much of the illicit caviar trade.[15]

SCOPE AND DIMENSIONS OF THE TRAFFIC

The lure of trafficking in wildlife is exemplified in the following quote by American Fish and Wildlife officer Tom Striegler: "[A] padded vest studded with 40 eggs from Australia's endangered black palm cockatoo, each worth $10,000, is far easier to smuggle than an equal-valued cache of cocaine, simply because custom officials aren't looking for cockatoo eggs."[16] In this quote is contained the nature of the problem—highly demanded, prohibited commodities precipitate a lucrative black market, made all the more profitable by ease of smuggling and the lax enforcement of regulations. Wildlife and products derived from certain animals are exceptionally valuable, with the price dependent on a number of variables including demand, rarity, luxury, and fashion status as well as real or perceived medicinal values.[17] To cite an extreme example, a rare macaw can fetch up to *$1 million* in Spain (see Table 3.1).[18]

The value of animals and wildlife products typically increases by 25%–50% as it moves through the supply chain—in some cases, the inflation is far greater. For example, an African gray parrot exported from the Ivory Coast increases from $20 at capture, to $100 at the point of export, to $600 for the importer in Europe or the United States, and to $1,100 to the specialist retailer. Local hunters earn little—perhaps $27 for a Melro bird, which then sells for $2,500 in Europe. At harvest, a hunter may receive $15 for a pink macaw, which will retail for $2,000 in Italy.[19] The Chinese three-striped turtle sells for $75 to $260 in Laos, $1,000 in Hong Kong, and is sold in the United States via the Internet for $2,200 per pair.[20]

The chain of the trade, from capture to the market, often requires intermediate processing, and may involve intermediate destinations. The international trade in particular is characterized by flexible distribution lines and networks of intermediaries, from harvesters through middlemen and on to wholesalers, exporters, processors, and retailers. The international trade can be viewed in terms of range states (exporters) in Central and Latin America, Asia, Eastern Europe, and Africa, and consumer states (importers) in North America, Western Europe, the Middle East, Japan, and Singapore. Some states are both range and consumer states, and include China, Australia, Canada, and South Africa. Routes used in the wildlife traffic are often complex, and take advantage of weaknesses and regulatory loopholes in the international trade control regime—especially in those cases in which controls have been relaxed to encourage free trade (as in the European Union), the traffic is made easier. A good example of the complexity of the traffic in wildlife is hummingbirds, illegally trapped in Latin America,

Table 3.1. Black Market Prices for a Few Selected Species

Species	Price
Rare orchid	$10,000 each
Rare cacti	$7,000 each
Trained falcon	$5,000 to $50,000 each
Rare macaw	$20,000 to $40,000 each
Tibetan antelope (wool)	$35,000 per shawl
Musk deer (musk)	$50,000 per kilo
Komodo dragon	$30,000 each
Rhino horn	$10,000 per kilo
Bear bile	$1,000 per 250cc
Tiger bone	$450 per kilo
Colophong beetle	$15,000 each
African elephant ivory	$750 per kilo

Sources: Francesca Colombo, "Animal Trafficking—A Cruel Billion Dollar Business," *Inter Press Service,* September 6, 2006, http://www.commondreams.org/headlines03/0906-06.htm (accessed September 24, 2010); Jane Holden, *By Hook or by Crook: A Reference Manual on Illegal Wildlife Trade and Prosecutions in the United Kingdom* (TRAFFIC International, 1998).

smuggled to Suriname, shipped to the Netherlands because of preferential trade agreements, and there sold to dealers who are free to travel without restrictions in the European Union.[21]

Traffickers often use the same routes as legal importers, including the use of direct flights and trans-Atlantic ships from range to consumer states—they triangulate routes, falsify certificates, and mix live illegal shipments of animals with legal exports. Common methods involve concealment, misdeclaration, permit fraud, and the laundering of wildlife products through complex reexport schemes. Many traders traffic protected wild species by falsifying paperwork to indicate specimens were artificially propagated or bred in captivity.[22]

Methods of concealing products for smuggling are limited only by relative bulk and the ingenuity of the smuggler. Exotic birds are squeezed into plastic tubes, while reptiles are packed into suitcases using paper or cloth bags stitched to the inside. Bird eggs, plants, and reptiles have been transported by concealing them in specially designed underwear. Reptile skins are hidden in shipments of cow hides, and ivory is dyed and concealed within timber shipments. Species have been transported using fake army and government plates, in ambulances, and inside wedding and funeral cars. Misdeclaration involves traffickers who make fraudulent claims on customs documents, a practice that typically includes the use of look-a-like species, the shipment of large quantities so as to facilitate inaccurate or false counts, listing the value of a shipment as much lower than it actually is, and claiming that an endangered species shipment has been artificially propagated or bred in captivity. Permit fraud involves wildlife traffickers who claim that legal documentation has been lost or stolen—sometimes permits are recycled, where dealers reuse the documentation to "launder" wild-caught animals as captive-bred. "Laundering" wildlife through reexport occurs when traffickers import the commodities through intermediate destinations for processing or manufacture—in such cases substituting legal for illegal products is made simpler. For example, reptile skins may be cut into sections, thus facilitating the insertion of many additional skins when the cut pieces are reexported to market.[23]

The traffic in wildlife and wildlife products is best understood when broken into categories, each with its own market, trafficking routes, and smuggling methods. Major categories include (1) the collecting of specimens by specialists; (2) the traffic in skins, furs, and fleeces; (3) the trade in traditional Asian medicines (TAMs); and (4) caviar smuggling (discussed fully in chapter 4). The most highly organized trade is in caviar (typically controlled by Russian organized crime groups); the least organized involves the market for specialists who collect specimens, and traditional Asian medicines and the traffic in skins, furs, and fleeces exhibit moderate levels of organization.[24]

Rare and exotic specimen trafficking is fueled by demand from collectors. The most prominent species in this sector of wildlife trafficking are tropical birds, reptiles, amphibians, and orchids, valued for their aesthetic appeal, breeding potential, and rarity. Collectors exist for all wildlife parts, dead specimens, insects, skulls, birds, and eggs. High prices are driven by the scarcity of the given specimen and the relative degree to which the specimen is regulated. There have been cases in which unscrupulous collectors intentionally destroyed the last habitats of plant species in the wild to prevent others from possessing them, resulting in extinction. With exotic species, networks are sometimes set up with a specific trade, or existing trade networks are used, often parallel to the legal trade. Knowledge of the legal market allows insiders to route specimens through countries with weak enforcement and monitoring protocols.[25]

The trade in furs and skins requires specialist skills involving identi-fication, capture, and killing. Untreated skins are exported to tanneries and then to manufacturers. The chain of production may involve several instances of import and reexport—allowing for opportunities to "launder" or insert prohibited skins into the trade. Ten million reptiles are killed annually for skins to make handbags and watch straps—while most of the trade is legal, illicit skins from the wild com-prise a significant but unknown portion of all skins sold. Skins and furs produce luxury clothes and accessories, a prime example being shah-toosh wool, produced from the under-fleeces of the Tibetan antelope (a critically endangered species). The shahtoosh fleeces are traded to India, where the legal trade in pashmina shawls (made from cashmere wool) is used for cover to transport shahtoosh. Finished shawls are sold in small quantities to tourists or shipped to consumer states, mostly in North America, Western Europe, Hong Kong, and Japan.[26] A single shawl can sell for as much as $35,000.[27] The traffic in shahtoosh may be increasing, fostered by the trade in tiger bones that are bartered across the Indian border for Nepalese and Tibetan wool.[28]

The trade in traditional Asian medicines (TAMs) has been practiced for 5,000 years. Although alternatives are available, the traffic is driven by beliefs (not always proven) concerning the medicinal properties of rare plants like orchids and ginseng, and products derived from animals like tigers, leopards, rhinos, bears, and musk deer.[29] The international trade in Chinese traditional medicines is expanding at a rate of 10% yearly—such demand coupled with a loss of habitat has reduced plant and animal populations, with 15%–20% of these species now endangered.[30] TAM ingredients are traded in raw form or in manufactured medicines such as pills or plasters. Raw ingredients are collected from Africa and the Americas, with most transported to Asia, especially China, where they are processed.[31] But China is not the only market: surveys of Chinatowns in New York City and San Francisco revealed significant percentages of traditional medicine shops selling products labeled as containing tiger bone, leopard bone, rhino horn, musk, and bear bile.[32] TAMs are trafficked through various routes, most prominently through legitimate pharmacies—this compounds enforcement problems because it is difficult to discern legal from illicit products. Prepared medicines are distributed to Chinese phar-macies and other outlets in East Asia and Western countries, but in some cases, tiger bones are shipped directly to pharmacies. Many seizures consist of small quantities smuggled in passenger luggage or in postal packages, but some larger commercial consignments are concealed in the shipment of legal goods.[33] The traffic in traditional Asian medicines and the skin, fur, and fleece trade have similar patterns of processing and distribution. There is some degree of organization in poaching to supply manufacturers—prosecutions in Western countries reveal that perpetrators are typically "legitimate" corporate traders (see Table 3.2).[34]

Although the traffic in wildlife is a global problem, some regions are especially at risk due to a variety of geographical, political, and socio-economic factors. A good example is the Russian Far East. For one, the region is especially rich in wild flora and fauna, containing both northern Siberian and south Manchurian species—poaching and smuggling involves approximately 160 different species, including Amur sturgeon, Far Eastern trepang, Asian black bear, musk deer, suppon, ginseng, Far Eastern leopard, Amur tiger, beluga whale, northern fur seal, and Pacific walrus. Another factor in the burgeoning illicit trade is that the far east of Russia is geographically proximate to East Asian countries, which has traditionally been a relatively large consumer of plant and animal products, particularly for use in traditional Asian medicines. Also, while household incomes in Russia have dropped, the economic well-being of East Asian countries has grown, precipitating an incentive for Russian wildlife suppliers to meet a rise in East Asian demand. In addition, the decline of the commercial fur industry in the Russian Far East has caused more than 90% of hunters in the region to supplement their income by illegally harvesting

Table 3.2. Animal Parts Used in Traditional Asian Medicines/ Symptoms Treated

Part Used	Symptom Treated
Rhino horn	High fever, heat stroke, erythema, purpura, vomiting of blood, nosebleed, convulsions, delirium, manic behavior
Tiger and leopard bone	Joint pain, paralysis, weak knees and legs, spasms, lower back pain, pain in bones
Bear gall	High fever and convulsions, spasms, hot skin lesions, red and swollen eyes, trauma, sprains, swelling and pain, hemorrhoids
Musk grains	Convulsions, delirium, stupor and fainting, closed disorders, titanic collapse, seizures, swelling and pain, toxic sores, carbuncles, coronary artery disease

Source: Leigh Henry, *A Tale of Two Cities: A Comparative Study of Traditional Chinese Medicine Markets in San Francisco and New York City* (TRAFFIC North America/WWF, 2004).

products such as musk, bear bile, velvet antlers, and ginseng. More-over, political, economic, and social changes in Russia beginning in the early 1990s have left a vacuum in enforcement, where environ-mental and customs authorities simply lack the capacity to regulate the illicit wildlife trade.[35] Much of the traffic involves organized and technologically well-equipped groups comprised of citizens from the Russian Federation, China, and Korea. Smuggling of live specimens and animal parts and derivatives is accomplished through a variety of means and transports: specially made, hermetically welded boxes are placed amid shipments of wood, timber, or scrap metal that hinders qualitative customs checks; products are hidden in containers with double bottoms; and specimens are concealed amongst active machi-nery like diesel locomotives or rail engines, where access is limited. Corruption among law enforcers and military personnel in Russian and Chinese border forces facilitate the traffic.[36]

THE ROLE OF ORGANIZED CRIME

Much of illegal wildlife trading involves small networks of people, including friends, associates, and family members. The large-scale legal trade is used as cover for illegal wildlife trafficking; in many cases, legal traders themselves become involved in the black market as they are best positioned to exploit regulatory weaknesses and trade routes. Still, depending on the particular commodity, the traffic in wildlife may exhibit a high degree of transnational organization. International wildlife trafficking networks have corrupted customs agents and other public officials, established transit hubs and networks of couriers, and developed procedures for laundering profits.[37] In some cases, existing organized crime groups have become involved, while some crime networks are extensive and highly sophisticated.[38] For example, in 1999 in Germany, a journalist who had mistakenly received a fax uncovered an international network of organized crimi-nals who were trafficking in a range of protected species. An investiga-tion revealed a group of Germans, Poles, Russians, and Indonesians committed more than 100 illicit trade offenses involving elephants trafficked from Indonesia to Argentina, China and Germany, tigers transported from Belgium to China and the United Kingdom, and komodo dragons smuggled from Indonesia to France and Mexico.[39] In the United Kingdom, a core of habitual offenders operate in organ-ized gangs—an analysis of persons convicted of wildlife crimes in Northumbria over a 12-month period showed that 50% of offenders had previous convictions for offenses such as drug trafficking, assault, burglary, and firearms violations.

While a 1999 report found that major organized crime group involvement in the Australian wildlife trade was minimal, other areas of significant organized crime activity have been well documented.[40] The Russian mafia and drug trafficking organizations from Latin America, Asia, and Europe use existing smuggling routes for drugs and small arms to trade in wildlife. In Brazil, a government commission documented a link between wildlife trafficking and the illicit trade in drugs and precious stones. In Mexico, numerous drug lords have been involved in wildlife trafficking, including Joaquin "El Chap" Guzman, who in 1993 had 70 protected species seized from his ranch.[41] Chinese Triads, including 14K and Wo Shing Wo, and organized African gangs are also involved.[42]

Authorities have observed a significant overlap between drug trafficking and the wildlife trade. The connection is multifaceted, and includes the parallel trafficking of drugs and wildlife along shared smuggling routes (where wildlife is a subsidiary trade), and the use of ostensibly legal shipments of wildlife to conceal drugs.[43] In Brazil, police estimate that 40% of all illegal drug shipments are combined with wildlife.[44] In Colombia, drug cartels operate in range areas for endangered tropical species, so a subsidiary trade in wildlife trafficking has naturally developed. The U.S. Fish and Wildlife Service (USFWS) reported that in 1993, a third of cocaine seized in the United States was associated with wildlife imports. In 1993, 41 boxes of 312 boas arrived in Miami with valid certification, but X-rays revealed condoms inside the snakes containing 120 kg of cocaine (most of the snakes died). The USFWS seized a total of $26 million of drugs in wildlife shipments that year, including heroin-filled condoms in the stomachs of goldfish. In 1995, 300 illegally exported turtles from Madagascar were shipped with 1.37 tons of marijuana. Other examples include customs agents at Heathrow Airport in London who found heroin packed into the shells of live snails, and investigators in Rome who discovered heroin smuggled in elephant tusks.[45]

Another dimension of the wildlife-drug trafficking nexus is the use of wildlife products as currency to "barter" for drugs, and the exchange of drugs for wildlife as part of the laundering of drug trafficking revenue. The USFWS says that loads of smuggled birds from Australia are exchanged for heroin in Bangkok, with the drugs flown back to Australia for resale. Drug smugglers have also been known to use dangerous wildlife such as tigers, venomous snakes, and crocodiles to protect drug shipments.[46]

Organized crime groups are also involved in the illicit elephant ivory trade. Multiton seizures recorded by the Elephant Trade Information System (ETIS) demonstrate the participation of highly organized criminal networks that exploit covert procurement and movement channels, use illicit proceeds to invest in facilities for storage and staging, exploit trading networks between range and end-user states, and corrupt regulators at seaports, airports, and border crossings. Indonesia, Thailand, Cambodia, and over one dozen African range states are most heavily implicated, while Nigeria seems to be most problematic when it comes to organized crime involvement in the ivory traffic.[47]

A CLOSER LOOK

BIRDS

Birds are perhaps the most sought-after creatures in the live animal trade.[48] Hundreds of thousands of wild birds are illegally smuggled every year, with approximately 250,000 illegally shipped into the United States. Endangered species such as salmon-crested cockatoos, Bali starlings, and red siskins are popular and rare—the illicit trade threatens their survival in the wild. There is a significant black market for finches, with goldfinches fetching 70 pounds sterling each—illegal trapping of endemic finches in Scotland is a significant problem. The illicit traffic in birds of prey has also grown in recent years. Political changes in Eastern Europe and the Commonwealth of Independent State (CIS) Republics precipitated an increase in smuggling of birds of prey, and with some species worth tens of thousands of pounds, organized international criminals have become involved. In 1991, a joint investigation by authorities in Denmark, Germany, and France discovered a smuggling ring based in Spain that operated throughout Europe, North America, and the Mediterranean. One case involved four Gyr falcons taken in Greenland and smuggled into France—their value was placed at $50,000 per animal.[49] In the United Kingdom, peregrine falcons are the most targeted species, although the theft of rare species such as golden eagles and ospreys occur as well. Law enforcement operations in the United Kingdom have uncovered gangs that demonstrate a high level of sophistication in the theft of birds of prey and their eggs.[50]

In Southeast and Central Europe, organized criminals hunt and smuggle song birds into northern Italy and Malta, where they are eaten as a delicacy. The illicit industry, thought to generate about €10 million annually, involves the shooting and exporting of hundreds of thousands of birds, many of them rare and/or protected by convention. In recent years, hunting hotspots have shifted from Hungary to Bulgaria, Romania, Serbia, and Montenegro, although the practice also occurs in Bosnia, Herzegovina, Macedonia, Albania, and Croatia. The main transit countries are Slovenia, Croatia, and Hungary. The red-breasted goose, corncrake, quail, and European turtle dove are all seriously threatened by illegal hunting practices. In 2003, Italian police seized a trailer carrying 120,700 specimens covering 83 different species—68 of the bird species were under permanent hunting prohibition, while 33 were rare species.[51]

Although most of the commercial exportation of wildlife from Mexico was banned in 1982, a significant illicit trade in parrots smuggled into the United States continued. As is typical of illegally trafficked wildlife, prices for parrots increase rapidly along the supply chain: a scarlet macaw may net $20 for the Mexican trapper, but retail in an American pet store for $4,000 (middlemen accrue much of the birds' final value). "Mules" (smugglers) typically move the birds across the U.S.-Mexico border by land, and receive $5–10 per bird.[52] In India, despite a total ban on trapping and trading since 1991, an illegal trade in wild birds flourishes, due mainly to the traditional practices of indigenous tribes that rely on trapping for their livelihoods. High demand is driven by a broad range of uses in the domestic market—aside from food, zoological, and medicinal purposes, birds are captured for release functions and for black magic and sorcery. A clandestine international traffic, mostly in parakeets and munias, flows through Nepal and Pakistan. In all, 36 CITES-listed species are traded in India.[53]

PLANTS

Rare orchid species are commonly harvested from the wild illegally. Commercial demand for orchids is so great that collectors pretend to be wardens to chase off competitors. In one case, 100 rare green-winged orchids were stolen from a nature preserve. In the United Kingdom alone, the collection of bluebell and snowdrop orchids is an illicit industry that generates one to three million pounds sterling per year—sometimes entire woods are stripped, with bluebell populations taking decades to recover. Collectors who seek out rare plant species have depleted whole regions of specimens, and are known to destroy wild plants to prevent rivals from securing them. Species such as the giant pitcher plant have been driven to near extinction. Declines due to overharvesting of wild plants such as orchids, bulbs, cacti, cycads, carnivorous plants, and airplants have resulted in many species being listed under the Convention on the International Trade in Endangered Species of Wild Fauna and Flora (CITES), the principal international convention that regulates the wildlife trade; and there is even a blanket ban on the export of specimens from Mexico. But enforcement is difficult, and high demand from collectors drives a black market and ensures widespread availability of CITES-listed specimens. In just one year alone in the United Kingdom, 12,000 plant specimens were seized because of permit issues. Smuggling is relatively easy, as identification of plant specimens requires the expertise of a specialist—a common technique is to claim plants collected in the wild were artificially propagated (recent advances in DNA testing can verify or disprove these claims).[54]

Approximately 50,000 to 70,000 medicinal and aromatic plants are used globally—4,000 of these are threatened, and 300 are listed in the CITES Appendices (many CITES-listed species are also used for their wood, and as ornamentation).[55] Plants have been a central component in traditional Asian medicines for thousands of years. In Cambodia, many residents rely on traditional medicine from plants (and animals) as their only source of health care[56]—out of the 12,807 traditional Chinese medicine sources, 11,146 (87%) are plants. In China, the traditional medicine industry accounts for approximately one-quarter of the country's pharmaceutical output, with major exports to Hong Kong, Japan, the United States, and countries throughout Southeast Asia. Unfortunately, many plant species are harvested beyond their regenerative capacity, resulting in serious declines and even localized extinctions.[57]

REPTILES

Approximately 500 reptile species are listed under CITES and afforded protections. Much of the reptile trade is legal, but the illicit trade, driven in part by specialist collectors, endangers many rare species in the wild. Illicit trafficking in live reptiles meets market demand for exotic pets—the European Union (EU) is one of the largest markets for live reptiles, with imports of live specimens increasing some 300% in the 1990s (Germany is perhaps the largest importer of live reptiles in the EU). The green iguana, royal python, and various geckos and chameleons dominate the trade, while Colombia, Madagascar, and El Salvador are the principle range states/suppliers.[58]

In addition to the exotic pet market, snakes, lizards, tortoises, and turtles are trafficked for their meat and for use in traditional medicines. At least 10 million reptiles are also killed annually for the reptile skin trade, a commercial enterprise that exploits species such as the spectacled caiman, water, and Nile monitor lizards, and the reticulated python. Most of the species used for the skin trade are listed under CITES (conversely, relative to birds and mammals, the live reptile trade is largely unregulated).

In recent years, reptile and amphibian exports have grown dramatically from range states such as Madagascar and the South Pacific islands. Skins are smuggled across borders in commercial shipments, while misdeclaration of the species involved, the number of skins, and the origin of the products are common. The U.S. Fish and Wildlife Service has uncovered reptile-smuggling rings in the Netherlands, Australia, Indonesia, and the United States (the Netherlands appears to be a hub for the illegal reptile trade globally). The illicit traffic in the United Kingdom has been linked to organized crime and drug smuggling.[59]

In China, reptiles are commonly used for food—much of the trade is illicit. A 2007 survey found that freshwater turtles and snakes accounted for the majority of the illegal wildlife food trade. Just one survey of Guangzhou markets found 5,000 to 24,500 turtles, three-quarters of which were listed as threatened species (70% of the 90 Asian freshwater turtle species are threatened).[60] Chinese demand for reptile meat and the depletion of once-abundant species on the mainland have precipitated international smuggling enterprises: snakes, freshwater turtles, and giant lizards are smuggled into China via the sea, while pangolin (a type of anteater) and giant lizards are moved over land from Vietnam and Laos. From sources in Myanmar to final market in China, the price of pangolin increases to 50 or 60 times its original value.[61] Elsewhere, 25 tons of freshwater turtles were exported every week from North Sumatra in 1999, but now the trade has declined as a result of overharvesting and declining populations. Rare and endemic species of Indonesian turtles are also popular in the exotic pet trade.[62]

Confusion endemic to reptile taxonomy is commonly used as a cover for misdeclaration of threatened species. Captive-bred specimens contribute a large percentage to market demand, but also make it relatively simple for illicit traffickers in wild-protected specimens to operate by commingling wild and captive-bred animals. Smuggling techniques are variable, and involve express mail shipments with false invoices and human couriers who hide animals in private luggage via air transport. Misdeclaration on customs documents is an oft-utilized method, while organized smuggling chains that employ anonymous couriers is increasingly common. Uncovering the trade in the importing country can be very difficult, unless, for example, there is consistent variation in the skins that denotes geographic origin.[63]

INVERTEBRATES

A legal and illegal traffic exists in rare butterflies, leeches, snails, spiders, scorpion beetles, and various insects. Birdwing butterflies from Southeast Asia can fetch up to $2,500 per pair. An illegal trade in live coral occurs on a large scale in the Philippines, with many reefs seriously damaged by overharvesting and the use of cyanide. Live coral and marine invertebrates are generally moved in commercial shipments or concealed with unregulated specimens, while spiders, scorpions, and insects are usually hidden in personal luggage.[64]

TIGERS

Tigers may become extinct outside of captive breeding, with fewer than 2,500 adult animals left in the wild.[65] The most significant factor driving the illicit harvesting of tigers and the traffic in tiger parts and derivatives is the market for traditional Asian medicines in China, the Republic of Korea, Japan, Malaysia, Vietnam, and Singapore (China instituted a domestic ban on the trade in 1993). Tiger meat is also considered a health tonic in China and Vietnam, and some Malaysian restaurants offer the meat as a luxury food. Tiger pelts are still used as traditional clothing in Tibet, and is a factor that drives the hunting of the animals in Nepal. While the domestic ban on tiger products and awareness campaigns have reduced the use of tiger parts as medicine in China, the trade in tonic wine derived from captive-bred tigers in "tiger farms" is likely to reduce the positive trend. In fact, extensive captive breeding and the failure to destroy existing stocks of tiger carcasses and parts may alter public perception and increase the demand for tiger products.[66] The increasing rarity of tigers has also increased the demand for other big cats, most notably various leopard subspecies. While tiger bone used as medicine has been largely eliminated in China, new markets for tiger and leopard skins have expanded, especially in Tibet—significant tiger and leopard skin seizures in Nepal and India since 2000 suggest a sizable illicit traffic and the presence of long-standing, highly organized smuggling operations.[67]

Three subspecies of tiger remain in Southeast Asia, but both endemic tiger subspecies from Indonesia are recognized as extinct. Poaching pressure remains high in some areas of Sumatra, Myanmar, and Malaysia.[68] Most tigers are harvested by professional hunters who sell directly to traders. In Indonesia, raw parts are exported before processing, but elsewhere, small family-owned processors develop traditional medicines locally. Retailers are typically traditional medicine businesses or restaurants that service urban centers. Black-market retailers operate in areas of weak enforcement—in North Sumatra, tiger parts are still sold openly.[69]

ORANGUTANS AND GIBBONS

Globally, two-thirds of the world's apes are on the World Conservation Union's "red list" of threatened species. Coupled with illegal logging and habitat loss, the hunting and capture of orangutans and gibbons in Indonesia pose a serious threat to those species. Most of the animals originate in Borneo and Sumatra and are transported to Bali and Java where they are widely kept as pets. Data from bird markets, wildlife rescue centers and zoos indicates that total loss to wild populations of targeted orangutan and gibbon species may amount to about 1% annually. In domestic markets, most live primates are trafficked for the pet market, but internationally, the trade is driven primarily by the biomedical research industry. Fortunately, the international traffic in primates has declined since the 1960s, though the trade is rather dynamic, with variation largely dependent on species type.[70]

ELEPHANTS

An increase in demand for ivory in the 1970s and 1980s due to rapid economic development in East Asia led to widespread poaching and drastic elephant declines in range states throughout Asia and Africa. In 1989, the 7th meeting of the Conference of Parties to CITES effectively banned all commercial international trade in elephants and elephant products, including ivory.[71]

Regulatory efforts at elephant conservation have been only partially successful. A large number of African and Asian range states rarely, if ever, report seizures of elephant ivory, and yet are implicated in large numbers of seizures elsewhere around the world. In Africa, 13 of 37 range states reported only 34 seizures from 1989 to 2009, but were implicated in well over 1,000 seizures internationally. The Democratic Republic of Congo reported six seizures but was implicated in 396 internationally, Angola reported zero but was implicated in 160, and Ghana reported two but was linked to 111 seizures. In Asia, neither Indonesia nor Cambodia reported any domestic ivory seizures, but were implicated in 51 and 26 seizures, respectively.

Nigeria and Thailand are especially problematic, based on an analysis from the Elephant Trade Information System (ETIS).[72] Nigeria did not report a single ivory seizure to ETIS from 1991 to 2009, yet ranks near the top of nations in the frequency and scale of the trade. Ivory seizures that implicate Nigeria are large in scale, indicating that organized crime and official corruption play a significant role in that country's illicit ivory trade.[73] Meanwhile, Thailand is exceeded only by China–Hong Kong SAR in the size of its illicit ivory market— driven largely by the tourist trade, the number of wild elephants in Thailand has been reduced to a high of 200,000 in the nineteenth century to no more than 2,000–3,000 today.[74]

ETIS characterizes 14 nations as having no effective law enforcement in the area of ivory trafficking and elephant conservation.[75] Three hundred and sixty-one tons of ivory have been seized globally and reported to ETIS since 1989; moreover, 34% of the total seized was shipped in large (greater than one ton) shipments, indicating the involvement of sophisticated criminal networks that possess greater levels of finance and investment as well as the ability to compromise law enforcement and customs officials.[76]

As of October 2009, the ETIS database contained 14,364 ivory seizure records (10% involved non-ivory elephant products) for the period 1989 to 2009. The database indicates ivory seizures in one form or another in 85 countries around the world, but collectively implicates 167 nations and territories in the trade. An ETIS analysis shows a decline in illicit ivory trading from a high in 1998 to a low point in 2003, a gradual increase from 2003 to 2007, and a sharper increase into 2009 (approaching 1998 highs).[77]

Significant progress has been made in the southern regions of the African continent, as robust wildlife management and law enforcement measures have led to an increase in elephant numbers in South Africa, Zimbabwe, Botswana, and Namibia. In other countries, the picture is not as good. Angola may be emerging as a significant country in the illicit ivory trade. A 2005 survey found 41 retail outlets near Luanda, Angola, selling ivory products, despite national laws prohibiting possession of ivory without proper documentation. Raw ivory appears to be acquired easily in the market, with significant quantities of carved objects imported illegally from the neighboring Congo Basin countries. A similar survey in Maputo, Mozambique, found 3,254 ivory objects displayed openly at 45 retail outlets. Moreover, 20% of the ivory objects were available in the duty-free area of the international airport, clearly demonstrating that they were taken out of the country in contravention of CITES regulations.[78]

REGULATION AND ENFORCEMENT

CITES is the principal international agreement among governments to combat wildlife trafficking. Protection is afforded in varying degrees to 33,000 species of animals and plants (not all endangered), including invertebrates, reptiles, mammals, fish, corals, birds, plants, and others. The text of the Convention was agreed upon in Washington D.C., in 1973 by 80 countries, and on July 1, 1975, it became law. CITES is legally binding on Parties to it, but does not replace national law—countries are expected to develop and implement domestic legislation to combat the problem. There are 172 Member Parties. Import, export, reexport, and introduction from the sea of listed species are subject to controls through licensing. Each Party to the Convention must designate one or more management authorities in charge of administering the licensing system and one or more scientific authorities to give advice on the effects of the trade on species. CITES does not have enforcement powers, and does not issue permits—management authorities in Party countries are responsible for regulation. Legally, import and export of CITES-listed species may occur only when proper documentation is obtained and displayed at ports of entry or exit.[79]

Under CITES, species are divided into three categories—Appendix I species are threatened with extinction, and trade is generally prohibited. Appendix II species are not necessarily endangered, but trade must be controlled to avoid endangerment. Appendix III species are protected in at least one country, and other member nations are asked to assist that country in protecting the listed species. Licensing and document requirements vary by the category to which the species belong. Fewer than 1,000 species belong to Appendix I—in fact, 95% of CITES-listed plants and animals are not endangered.[80] An import permit (from the importing country's management authority) and an export permit (from the exporting country's management authority) are required for each specimen. These are to be issued only when it does not endanger the species, the specimen is not to be used for *primarily* commercial purposes, and, in the case of live specimens, the scientific authority is satisfied that the recipient is able to care for the plant or animal. Appendix II specimens do not require an import document, unless required by the import country.[81]

CITES is a complex treaty that has become more complicated over the years as numerous resolutions that pertain to interpretation, definition, and application of regulations have been adopted. In 1994, CITES Parties agreed on additional measures focusing on the protection of tigers, specifically acknowledging the threat posed by the use of tiger parts for the traditional medicines market. In 1997, Parties to the Convention also acknowledged the need for additional measures to foster bear conservation and to thwart the illicit bear trade.[82] CITES has implemented additional species-specific programs for certain animals and plants, including elephants, falcons, great apes, hawksbill turtles, mahogany, and sturgeon. In recent years, MIKE (Monitoring the Illegal Killing of Elephants) and ETIS (Elephant Trade Information System) have emerged to bolster CITES in elephant conservation.[83]

In general, there is a lack of international coordination in the monitoring and enforcement of wildlife trafficking. INTERPOL has a Wildlife Crime Subgroup to facilitate international cooperation. However, collaboration on an international scale is beyond the resources and

mandate of the CITES Secretariat. CITES lacks the authority to significantly curb illicit wildlife trafficking—management authorities
within nations are expected to provide for inspection resources, but
many signatories are unwilling to provide funds. CITES does not even
place a requirement on member states to provide enforcement capabilities, and has no enforcement power of its own.[84]

Difficulties with the implementation of CITES regulations are
quite pronounced in the EU. Since the liberalization of trade policies
and the advent of the Single Market in 1993, a series of legislative
and administrative changes significantly altered the regulation of the
wildlife trade in the EU. Regulation 3626/82 applied CITES to all
EU member states and actually imposed stricter rules than the Convention, but it also largely removed the provisions for monitoring
trade across borders shared by EU member states. Now thousands of
fewer customs officials police internal EU borders, and since 1993
nations are far less likely to inspect wildlife shipments—wildlife traders find it relatively easy to circumvent regulations by simply importing species into neighboring countries instead of directly to the
destination state. Moreover, Authorities in the first point of entry typically do not have access to the original import permit since this usually remains in the destination state—obviously without the original
import documents, CITES controls cannot be applied.

Regulation 3626/82 was inherently problematic not only because it
made it far easier to move goods (legal and illicit) across internal
borders, but also because it failed to specify certain aspects of enforcement and implementation integral to the EU-wide application of
CITES. Lack of specificity and vague language provided individual
member states with broad latitude to interpret rules, precipitating a
range of national measures incompatible with the common market.
The text of Regulation 3626/82 failed to adequately define the obligations of member states regarding CITES, and did not outline powers
of enforcement or establish applicable penalties for Convention violators. In addition, the regulation was not flexible enough to allow for
changes to the lists of CITES species, while some member states were
unable to seize or apply penalties when illicitly traded specimens were
discovered in transit.[85]

Aside from the problems with Regulation 3626/82, some EU states
simply failed to develop legislation to implement CITES and EU regulations. For example, Spain uses customs contraband legislation,
which makes no specific reference to CITES, and actually creates
obstacles for the enforcement and prosecution of CITES violations.[86]
Additionally, in some member states, the scientific authorities do not
possess the expertise to advise management authorities on biology

and conservation issues, while most management authorities cite the need for additional resources to implement CITES in their countries.[87] The complexities and frustrations of CITES implementation in the EU is exemplified by the following anecdote:

In April 1994, two bird shipments involving 240 hummingbirds arrived from Peru in Belgium, destined for a trader in the Netherlands. One of the shipments was accompanied by an expired Peruvian export permit. The shipments were controlled by veterinary inspectors in Belgium, who did not realize the Peruvian permit had expired. The Belgian veterinary inspector trusted that a more thorough inspection of the shipment would be made in the Netherlands. Neither shipment, however, was inspected by officials in the Netherlands. On 11 July 1994, a third shipment of hummingbirds, destined for the same Dutch trader, arrived in Belgium. The shipment was accompanied by a copy of a valid Peruvian export permit and an invoice. The Dutch trader presented three Dutch import permits to Belgian Customs officials who noticed discrepancies in the numbers and species of birds (all Appendix II) listed on the documents. The Dutch permits made no reference to the number of the Peruvian export permit. In light of the two previous problematic hummingbird imports in April 1994, Belgian authorities informed Dutch authorities of this importation, and Dutch authorities advised the Belgians to confiscate the birds, reasoning that the shipment was accompanied by a copy and not an original export permit (under Dutch legislation, a Dutch import permit is only considered valid when matched with an original valid export permit). Since the fact that the shipment was accompanied by a copy of a valid export permit did not constitute an infraction of EU CITES Regulations, Belgian Customs officials informed Dutch authorities that they were unable to confiscate the birds on these grounds. Also, in Belgium, the fact that the Dutch permits failed to list the number of the Peruvian export permit is considered an administrative error on the part of the Dutch Management Authority, and confiscating specimens for this reason would punish the trader and pose the problem of housing the animals. For these reasons, Belgian and Dutch authorities agreed to an arrangement which would allow Dutch Customs officials to confiscate the birds in the Netherlands. The shipment was officially sealed, issued a T1 transit document, and sent to the Hazeldonk Customs post to be inspected by Dutch officials. By the time the hummingbirds arrived in Holland later that evening,

however, the Dutch Customs official with whom the arrangements had been made had finished his shift, and the new attend-
ing official had not been informed of the impending arrival.
When the shipment arrived, the Dutch trader collected the birds.
The following day upon learning of the events, the General
Inspection Service visited the trader's premises. The trader stated
to the inspectors that all birds were dead on arrival, and that he
had incinerated the bodies. Although they suspected that the trader
had sold or otherwise distributed the birds in the Netherlands,
without physical evidence, there was nothing further Dutch
officials could do.[88]

In June 1997, Regulations 3626/82 and 3418/83 were replaced with
the current EU Wildlife Trade Regulations. The new regulations
covered imports and exports of wildlife and wildlife products into
and out of the EU, and additionally addressed trade between and
within member states. All CITES provisions were incorporated along
with some more stringent regulations. Species are listed in four
annexes, with Annex A and Annex B categories roughly equivalent to
Appendix I and Appendix II CITES lists. Import permits are required
only at the first point of entry in the EU for endangered and threatened species (Annexes A and B), while *notification* permits are mandated for other specimens. Permits are not required for trade
between member states.[89]

One country with relatively rigorous controls is the United Kingdom,
where wildlife trafficking, in addition to CITES and the EU Regulations, is covered by the Customs and Excise Management Act of 1979
(CEMA), and the Control of Trade in Endangered Species regulations
(COTES). CEMA provides for a maximum seven-year sentence,
COTES has a two-year sentence maximum, but both have potential
unlimited fines. Nevertheless, in practice the penalties handed out
under CEMA and COTES have been quite low. No judge has ever
imposed a custodial sentence for a wildlife trade crime under COTES.
In fact, offenses committed under COTES are not even subject to
arrest. CEMA does provide for arrest, but no judge has ever imposed
the maximum seven-year sentence. Still, other UK provisions suggest
the nation may be taking the problem of wildlife trafficking seriously.
Most police forces in the United Kingdom have wildlife liaison officers. Also, the Partnership for Action Against Wildlife Crime brings
together statutory and non-statutory agencies, an effort that includes
the National Wildlife Crime Intelligence Unit.[90]

In the United States, CITES is implemented through the Endangered Species Act (ESA), while illegal wildlife traffickers can be

prosecuted under both the ESA and the Lacey Act. In the United States, the maximum penalty for individuals convicted of wildlife crimes is five years in prison and a $250,000 fine. However, the ESA only prohibits the import and export of, and interstate commerce in, listed species, while neglecting to explicitly prohibit intrastate sales. Investigation and enforcement functions are typically shared by the U.S. Fish and Wildlife Service and Customs, but numerous departments can be involved, including Homeland Security, Justice, Agriculture, Commerce, State, and Interior. Resources devoted to wildlife trafficking in the United States have been relatively low—in the 1990s, only two dozen wildlife officers patrolled the entire U.S.-Mexico border. One promising development occurred on September 22, 2005, when the U.S. Department of State announced the formation of the Coalition Against Wildlife Trafficking (CAWT), initially comprised of the groups Conservation International, the Save the Tiger Fund, the Smithsonian Institution, TRAFFIC International, Wild Aid, the Wildlife Conservation Society, and the American Forest and Paper Association. CAWT aims to support the Association of Southeast Asian Nations (ASEAN) to develop a regional action plan to combat wildlife trafficking.[91]

A successful enforcement initiative in the United States involved the closing of a regulatory loophole in the ESA and the Rhino and Tiger Conservation Act, both of which had failed to effectively control the trade in traditional Chinese medicines. The problem with the ESA arose from the fact that the U.S. government shouldered the burden of proof to demonstrate that medicines *claiming* to contain components of protected species actually did—however, forensic techniques are often unable to confirm the presence of ingredients such as rhino horn or tiger bone in medicines *labeled* as containing them. TRAFFIC International, the world's largest wildlife trade monitoring network (TRAFFIC works closely with the CITES Secretariat), persuaded the U.S. Congress to enact the Rhino and Tiger Product Labeling Act (RPTLA), which "prohibits the import, export and sale of any product for human consumption or application containing, *or labeled or advertised to contain* [emphasis added], any substance derived from any species or rhinoceros or tiger." The RPTLA provides for maximum penalties of six months in jail and a $12,000 fine per violation.[92] The legislation was successful, in part—surveys of traditional Chinese medicine shops in San Francisco and New York City from 1996 to 2003 found a significant decline in the percentage of stores claiming to sell products containing tiger bone and rhino horn. Unfortunately, there appears to have been a substitution effect, as the percentage of stores offering leopard bone increased (as most of the products labeled as containing leopard bone had previously used tiger bone, suppliers evidently met customer demand by substituting a product reputed to provide similar effects).[93]

The case of the RPTLA provides important lessons for regulators of wildlife trafficking. First, even thoughtfully developed regulations can precipitate unintended (if not unforeseen) consequences. While limiting or prohibiting the traffic in endangered and threatened species is clearly desirable, limiting and prohibiting *any* commodity that is highly demanded precipitates lucrative black markets. In the case of wildlife trafficking, there is not one market, but many markets for a broad range of products. Complex and varying international regulations can even exacerbate the illicit traffic in some species, as when suppliers meet demand by substituting products. Clearly, some efforts, though well-meaning, could be counterproductive. For example, listing species under CITES that are not endangered may increase the demand and value of those species, thus inadvertently precipitating an increase in the illicit traffic.

Perhaps the most significant issue in wildlife trade regulation is the lack of consistent and severe penalties for illicit traffickers. Mexico is one country with relative harsh penalties. Prison time there can range from six months to six years—in 2002, 17 people were indicted and paid fines totaling $580,000. Traffickers may receive a maximum of five years in Spain, and two years in Italy (but 12 if they are connected to the Mafia). But in Brazil, even significant perpetrators may post a $100 bond and perform community service with no jail time. Naturally, when fines and penalties are insignificant compared to the illicit profits, the deterrent effect of law enforcement is negated. A good example is the Renaissance Corporation, which processed 138 shawls from 1,000 endangered Tibetan antelopes with a value of 353,000 pounds sterling each. When caught, the company was fined just 1,500 pounds sterling (in fact, the *maximum* penalty provided for in this case was only 5,000 pounds sterling).[94]

A promising development in the fight against wildlife trafficking is the increasing use of forensic science, especially DNA analysis, in wildlife crime investigations. DNA paternity analysis is used to disprove claims of captive breeding, while even small fragments of fur, feathers, or skin at a perpetrator's home can be linked to the same materials at a crime scene. In cases where the illegal trade of an endangered species is disguised as the lawful trade of closely related species, DNA analysis allows for the precise identification of the endangered species and its derivative products. In 1993, the first DNA tests were carried out on birds of prey in captivity—of the 514 peregrines and goshawks declared as captive bred, 10% were not related to their parents, clearly indicating the unrelated birds had been mixed in from wild capture. In the year following the testing, claims of captive breeding of peregrines and goshawks fell by 20%.[95]

SUMMARY AND CONCLUSIONS

The illicit traffic in wildlife is a significant transnational industry fueled by demand for these commodities. While degree of organization, method of smuggling, and profitability vary by product, the trade generally features large profit margins—a characteristic driven by high demand, the rarity of many species, the ease of smuggling, and a lax and uncoordinated international enforcement effort. The participation of significant and established transnational crime groups has been well documented. The environmental consequences of the illegal trade in exotic plants and animals are significant, and potentially include the extinction of species, a reduction in biodiversity, and the collapse of some ecosystems. Human impacts include the corruption of public officials, violence directed against law enforcers, and the spread of disease.

A reduction in biological diversity is also linked to economic factors, and has an impact on economic relationships and the economies of individual nations. A reduction in biodiversity even has global economic consequences, as potential applications in biotechnology, biomedicine, pharmaceutical, and food industries are diminished. In addition to disrupting the natural processes that regulate the stable functioning of entire ecosystems, the loss of biodiversity produces significant negative outcomes related to the ethical, social, aesthetic, and religious dimensions of society.[96]

The international effort to curb the illicit wildlife trade has been generally poor. Bans of some products in the European Union are a hopeful sign, but consistency and cooperation on a global scale is required. CITES in particular exemplifies the feeble quality of the international effort, as it has little real authority to bind Parties to Convention protocols. While laws and regulations can provide effective mechanisms for control, law enforcement and appropriate ethical governance are the critical components of any control effort—public corruption associated with the wildlife trade must be reduced. In any event, the adoption of serious and consistent penalties across both range and consumer states would be one relatively effective way to deter illegal wildlife trafficking.

Wildlife trade chains are highly complex and variable, with different market conditions and incentives from harvest to end consumer. Interventions that focus on particular sectors of supply chains, some of which span thousands of miles and cross several international borders, may not be successful in limiting the illicit wildlife trade overall. The assumption that poverty is a major driver in the supply of regulated wildlife may be incorrect—surveyed experts believe that increasing income and diversifying livelihoods in rural communities probably has a relatively small impact on illegal wildlife harvesting. Many harvesters are not poor, and those who are poor do not drive the wildlife trade. Conversely, increasing affluence and disposable income in consumer states are directly related to increasing demand for wildlife products in particular regions. Both harvesters and traffickers are highly responsive to market opportunities presented by the regulation of selected species, and display mobility between markets, locations, and products in order to satisfy demand. Factors related to infrastructure development and trade expansion increase the supply of wildlife products in regions that are developing economically. Specifically, improved communications, roads, and increased access to remote areas through illegal logging facilitates harvesting and extraction of wildlife.[97]

Experts examining the wildlife trade in Cambodia, Laos, Vietnam, and Indonesia observed that aside from formal law enforcement, traditional practices, customary norms, tenure arrangements, and voluntary agreements have been highly effective in limiting the illicit trade, but that little attention has been paid to such initiatives. Resource management practices can be partially successful in reducing the illegal wildlife trade. Harvest controls, species management plans, closed seasons, and limits on technology are helpful only to the extent that the complex factors that influence the sustainability of harvesting regimes (and what level of exploitation is sustainable in a particular case) is fully understood.[98] Though underused, price- and market-based instruments show promise in reducing illicit trafficking—buying agreements, tax incentives, product certification, and price controls may be effective deterrents, especially when implemented across a range of participant groups.[99]

Raising awareness could be a critical factor in reducing the illegal trade of wildlife. For example, efforts to reduce the use of tiger bone and rhino horn in traditional Chinese medicines have been far more successful in San Francisco than New York City, likely because of far greater outreach and education efforts in California. Awareness campaigns seem to be more effective among consumers than either harvesters or traders, though research indicates the effects may be short-lived.

The role of nongovernmental organizations (NGOs) and private foundations in the fight against wildlife trafficking is significant. TRAFFIC (a global monitoring network for wildlife trafficking), the World Wide Fund for Nature (WWF), the World Conservation

Union (IUCN) and the Rufford Maurice Lang Foundation are prime examples of nongovernmental entities that monitor and investigate unsustainable wildlife trading and illicit trafficking, and spur governments to action.[100]

That overharvesting of some species and unsustainable practices are not defined as illegal because of economic interests was evident at the 2010 CITES conference in Qatar, where delegates failed to enact protections for bluefin tuna or red and pink corals, and declined to tighten controls on the domestic trading of tiger parts and products from tiger "farms."[101] Clearly, the legal harvesting and trade of threatened species is driven and shaped by consumer demand and the natural political tension between commercial and environmental interests. A broad perspective that seeks solutions requires that the problem of species endangerment be not viewed only as a product of illicit trafficking.

NOTES

1. *What's Driving the Wildlife Trade? A Review of Expert Opinion on Economic and Social Drivers of the Wildlife Trade and Trade Control Efforts in Cambodia, Indonesia, Lao PDR, and Vietnam* (Washington, DC: East Asia and Pacific Region Sustainable Development Department, World Bank, 2008).

2. Donovan Webster, "The Looting and Smuggling and Fencing and Hoarding of Impossibly Precious, Feathered, and Scaly Wild Things," *Trends in Organized Crime* 3, no. 2 (1997): 9–10.

3. World Wildlife Fund, http://www.worldwildlife.org (accessed September 17, 2010); Havocscope, http://havocscope.com/trafficking/wildlife.htm (accessed September 17, 2010).

4. Havocscope, http://havocscope.com/trafficking/wildlife.htm (accessed September 17, 2010).

5. Jane Holden, *By Hook or by Crook: A Reference Manual on Illegal Wildlife Trade and Prosecutions in the United Kingdom* (TRAFFIC International, 1998).

6. http://www.worldwildlife.org; http://www.traffic.org/generaltopics.

7. Ben Davies, *Black Market: Inside the Endangered Species Trade in Asia* (San Rafael, CA: Earth Aware Editions, 2005); Alan Green, *Animal Underworld: Inside America's Black Market for Rare and Exotic Species* (New York: Public Affairs, 1999).

8. Havocscope, http://havocscope.com/trafficking/wildlife.htm (accessed September 17, 2010); Francesca Colombo, "Animal Trafficking—A Cruel Billion Dollar Business," *Inter Press Service*, September 6, 2006, http://www.commondreams.org/headlines03/0906-06.htm (accessed September 24, 2010); "Organized Crime Fuels Illegal Ivory Surge in Africa," http://www.panda.org (accessed September 24, 2010).

9. Webster, "The Looting and Smuggling and Fencing and Hoarding."

10. Colombo, "Animal Trafficking."

11. Webster, "The Looting and Smuggling and Fencing and Hoarding."

12. "Organized Crime and the Environment," *Trends in Organized Crime* 3, no. 2 (1997): 4.

13. Adrian Levy and Cathy Scott-Clark, "Poaching for bin Laden," *Guardian*, May 5, 2007, http://www.guardian.co.uk/world/2007/may/05/terrorism.animalwelfare (accessed November 1, 2010).

14. Dee Cook, Martin Roberts, and Jason Lowther, *The International Wildlife Trade and Organised Crime: A Review of the Evidence and the Role of the UK* (Regional Research Institute, University of Wolverhampton, 2002).

15. "Switching Channels: Wildlife Trade Routes into Europe and the UK" WWF/TRAFFIC, December 2002, http://www.wwf.org.uk/filelibrary/pdf/switchingchannels.pdf (accessed November 1, 2010); "United States International Crime Threat Assessment," *Trends in Organized Crime* 5, no. 4 (2000): 56–59; "Organized Gangs Move into Wildlife Trafficking," June 17, 2002, http://www.wwf.org.uk/news/n_0000000589.asp.

16. Webster, "The Looting and Smuggling and Fencing and Hoarding."

17. Levy and Clark, "Poaching for bin Laden."

18. Colombo, "Animal Trafficking."

19. Ibid.

20. "Just the Facts," Royal Canadian Mounted Police, *Gazette* 66 (2004).

21. Colombo, "Animal Trafficking"; Cook et al., *The International Wildlife Trade and Organised Crime*.

22. Holden, *By Hook or by Crook*.

23. Cook et al., *The International Wildlife Trade and Organised Crime*, 19–20.

24. Ibid., 9–10.

25. Ibid., 10–11.

26. S. R. Galster, S. F. LaBudde and C. Stark, *Crimes against Nature: Organized Crime and the Illegal Wildlife Trade* (The Endangered Species Project, 2004).

27. Holden, *By Hook or by Crook*.

28. Galster et al., *Crimes against Nature*.

29. Cook et al., *The International Wildlife Trade and Organised Crime*.

30. Hongfa, Xu, and Craig Kirkpatrick, *The State of Wildlife Trade in China: Information on the Trade in Wild Animals and Plants in China 2007* (TRAFFIC East Asia, 2007).

31. Cook et al., *The International Wildlife Trade and Organised Crime*.

32. Leigh Henry, *A Tale of Two Cities: A Comparative Study of Traditional Chinese Medicine Markets in San Francisco and New York City* (TRAFFIC North America/WWF, 2004).

33. Cook et al., *The International Wildlife Trade and Organised Crime*, 12–13.

34. Ibid.

35. Sergey N. Lyapustin, Alexey L. Vaisman and Pavel V. Fomenko, *Wildlife Trade in the Russian Far East: An Overview* (TRAFFIC Europe-Russia, 2007).

36. Ibid.

37. Holden, *By Hook or by Crook*.

38. "Organized Gangs Move into Wildlife Trafficking," http://wwf .org.uk/news/n_0000000589.asp; Colombo, "Animal Trafficking"; Holden, *By Hook or by Crook*.

39. Cook et al., *The International Wildlife Trade and Organised Crime*.

40. Galster et al., *Crimes against Nature*; "Organized Gangs Move into Wildlife Trafficking"; Cook et al., *The International Wildlife Trade and Organised Crime*.

41. Colombo, "Animal Trafficking."

42. Ibid.

43. Cook et al., *The International Wildlife Trade and Organised Crime*.

44. "Smuggling's Wild Side in Brazil: Animal Trafficking Sucks the Life from the Amazon Rainforest," *Washington Post*, December 9, 2001, A38, http://latinamericanstudies.org/brazil/smuggling.htm (accessed November 1, 2010).

45. Cook et al., *The International Wildlife Trade and Organised Crime*, 14.

46. Ibid.

47. Tom Milliken, R. W. Burn and L. Sangalakula, *The Elephant Trade Information System (ETIS) and the Illicit Trade in Ivory* (TRAFFIC East/ Southern Africa, 2009).

48. Stephen V. Nash, *Sold for a Song: The Trade in Southeast Asian non-CITES Birds* (TRAFFIC Southeast Asia, 1993).

49. Holden, *By Hook or by Crook*.

50. Ibid.

51. *The Illegal Trade in Wild Birds for Food through Southeast and Central Europe* (TRAFFIC, 2008).

52. Jose Gobbi, Debra Rose, Gina De Ferrari and Leonora Sheeline, Parrot Smuggling Across the Texas-Mexico Border (TRAFFIC USA, 1996).

53. Ahmed Abrar, *Live Bird Trade in Northern India* (TRAFFIC India, 1997).

54. Holden, *By Hook or by Crook*.

55. Katalin Kecse-Nagy, Dorottya Papp, Amelie Knapp, Amelie, and Stephanie Von Meibom, *Wildlife Trade in Central and Eastern Europe: A Review of CITES Implementation in 15 Countries* (TRAFFIC Europe, 2006).

56. David Ashwell and Naomi Walston, *An Overview of the Use and Trade of Plants and Animals in Traditional Medicine Systems in Cambodia* (TRAFFIC Southeast Asia, 2008).

57. Hongfa and Kirkpatrick, *The State of Wildlife Trade in China*.

58. Mark Auliya, *Hot Trade in Cool Creatures: A Review of the Live Reptile Trade in the European Union in the 1990s with a Focus on Germany* (TRAFFIC Europe, 2003).

59. Holden, *By Hook or by Crook*.

60. Ibid.

61. Hongfa, 2007.

62. *What's Driving the Wildlife Trade?*

63. Auliya, *Hot Trade in Cool Creatures*.

64. Holden, *By Hook or by Crook*.

65. Kristin Nowell and Xu Ling, *Taming the Tiger: China's Markets for Wild and Captive Tiger Products since the 1993 Domestic Trade Ban* (TRAFFIC East Asia, 2007).

66. *What's Driving the Wildlife Trade?*

67. Nowell and Ling, *Taming the Tiger*.

68. What's *Driving the Wildlife Trade?*

69. Julia Ng and Nemora, *Tiger Trade Revisited in Sumatra, Indonesia* (TRAFFIC South East Asia, 2007); *What's Driving the Wildlife Trade?*

70. Vincent Nijman, *In Full Swing: An Assessment of Trade in Orang-utans and Gibbons on Java and Bali, Indonesia* (TRAFFIC Southeast Asia, 2005).

71. Daniel Stiles, *The Elephant and Ivory Trade in Thailand* (TRAFFIC Southeast Asia, 2009).

72. Milliken et al., *The Elephant Trade Information System*.

73. Ibid.

74. Stiles, *The Elephant and Ivory Trade in Thailand*.

75. Milliken et al., *The Elephant Trade Information System*.

76. Ibid.

77. Ibid.

78. Tom Milliken, Alistair Pole, and Abias Huongo, *No Peace for Elephants: Unregulated Domestic Ivory Markets in Angola and Mozambique* (TRAFFIC East/Southern Africa, 2006).

79. "What is CITES?" at http://www.cites.org/eng/disc/what.shtml (accessed September 24, 2010).

80. Ibid.

81. Ibid.

82. Henry, *A Tale of Two Cities*.

83. Ibid.

84. Cook et al., *The International Wildlife Trade and Organised Crime*, 28–30; CITES Web site, http://www.cites.org (accessed September 24, 2010); Karin Berlchoudt, *Focus on EU Enlargement and the Wildlife Trade: a Review of CITES Implementation in Candidate Countries* (TRAFFIC Europe, 2002).

85. Elizabeth H. Fleming, *The Implementation and Enforcement of CITES in the European Union* (TRAFFIC Europe, 1994).

86. Ibid.

87. Ibid.

88. Ibid., 20–21.

89. Holden, *By Hook or by Crook*.

90. Cook et al., *The International Wildlife Trade and Organised Crime*, 26–30; Holden, *By Hook or by Crook*.

91. "United States Forms Global Coalition Against Wildlife Trafficking," September 23, 2005, http://www.america.gov/st/washfile-english/2005/September/20050923155154lcnirellep0.4162409 (accessed November 1, 2010).

92. Henry, *A Tale of Two Cities*, 3.

93. Henry, *A Tale of Two Cities*.

94. Colombo, "Animal Trafficking."

95. Holden, *By Hook or by Crook*.

96. Lyapustin et al., *Wildlife Trade in the Russian Far East*.

97. *What's Driving the Wildlife Trade?*

98. Ibid.

99. Ibid.

100. Ibid.

101. "Salamander Protected from Trade," *Tribune Review*, March 22, 2010, A3.

CHAPTER 4

Illegal Fishing

Illegal, unreported, and unregulated (IUU) fishing is a global problem that threatens a broad range of fish species and fishing stocks. The United Nations' Food and Agriculture Organization (FAO) concluded that approximately 75% of the world's fish stocks are fully exploited, overexploited, or depleted, while IUU fishing may amount to a third of total world catches. The Commission for the Conservation of Antarctic Marine Living Resources (CCAMLR) estimated that 16.5% of the total catch of Patagonian toothfish in 2003–2004 was illegal, the International Commission for the Conservation of Atlantic Tuna (ICCAT) found that 25,000 tons (18%) of all fishing for tuna in 2001–2002 was likely attributable to illicit fishing, and the North East Atlantic Fisheries Commission (NEFC) reported that 27% of redfish caught in 2002 was landed by IUU ships. In some important fisheries, illicit activity accounts for perhaps 30% of total catches, and in certain ports, an estimated 50% of all fish are harvested illegally. Coupled with an overcapacity of legitimate fishing trawlers, illicit fishing practices have precipitated the collapse or near collapse of some fish populations. Moreover, the negative consequences of illicit fishing are not merely environmental, but include a range of deleterious economic and social (human) outcomes.[1] IUU fishing has a negative impact on broader marine ecosystems, and causes substantial damage to the food security and livelihood of coastal populations in developing countries. While it is difficult to gauge, IUU fishing may cost developing countries between $2 billion and $15 billion annually.[2]

Illegal, unreported, and unregulated fishing practices are variations of the same problem—the unsustainable harvesting of fish stocks. Offenses include fishing without a license or out of season, harvesting prohibited species, using banned fishing gear, overfishing, and failure to report or misreporting catch weights.[3] Another illicit practice is the dumping of fish and "high grading" (retaining only larger, higher-quality fish), which depletes the fishable stock and skews stock assessments integral to sound management.[4] Internationally, fishing in a state's jurisdictional waters without authorization and fishing in contravention of Regional Fishing Management Organizations (RFMOs) are also common infractions.[5] Most illegal fishing in the South African Development Community Region (SADC) is perpetrated by Distant Water Fishing Nations (DWFNs) such as China, Indonesia, Russia, Korea, Spain, and Taiwan — large foreign commercial interests essentially rob developing nations of their fish resources.[6]

For the purpose of clarity, a full definition of IUU fishing is obligatory.*

Illegal Fishing—where vessels operate in violation of the laws of a fishery. This can entail fishing with no license at all, or fishing in contravention of the terms of the license, for example by using outlawed fishing gear. This definition is used both for fisheries that are under the jurisdiction of a coastal State, and for those that are regulated by Regional Fisheries Management Organizations (RFMOs).

Unreported fishing—fishing that has been unreported or misreported to the relevant national authority or regional fisheries management organization, in contravention of national and international laws and regulations.

Unregulated fishing—this generally refers to fishing that is conducted by vessels without nationality, or vessels flying the flag of a State not party to the regional organization governing the particular fishing region or species. Unregulated fishing can also relate to fishing in areas or for fish stocks for which there is a lack of detailed knowledge of the resource, and therefore no conservation or management resources in place. In both these cases, vessels must be fishing in a manner that violates the conservation and management measures of the regional organization, and/or international law, to warrant inclusion under the term "unregulated fishing."[7]

There exist significant incentives and minimal disincentives to fish illegally. Legal and illegal fish are sold on the same markets, but legitimate fishers pay higher operating costs associated with licensing and overhead due to conservation and management measures. A good

---

*In this chapter, the terms "IUU fishing" and "illicit fishing" will be used interchangeably.

example is longline vessels that set their lines to minimize by-catch (catch of non-targeted species such as sea birds, turtles, and sharks) in accordance with regulations, but are then placed at a competitive disadvantage relative to longliners that fish with little concern for non-targeted species.[8] Illegal operators don't have to pay for licenses, observers, vessel monitoring systems, or catch documentation schemes, and further increase their profits by exploiting workers in low-wage countries. Moreover, the unfair competition from illicit operations may pressure legitimate outfits to cheat as well, thus precipitating a snowball effect where unsustainable practices multiply.[9]

A major incentive to exceed quotas or otherwise engage in IUU fishing is the exceptional value of some target species. A single Patagonian toothfish may be worth $1,000; a tuna, $50,000, even $100,000. Illegal fishing is so lucrative that the net profits of a vessel from a single excursion often exceed the price of the ship itself (so abandoning a seized ship may reflect no serious loss). Another major issue is the difficulty of enforcement—the world's oceans are vast, the resources devoted to regulatory compliance too meager, and the overall chances of being caught minimal. Even when the slim chances of apprehension are overcome, penalties typically involve little or no jail time, and fines are dwarfed by illicit profits.[10] Naturally, as demand for fish remains high and restrictions on commercial fishing increase, IUU harvesting will continue. In 2006, the National Oceanic and Atmospheric Administration opened 750 investigations into illegal fishing in the northeastern United States, an increase of 108% over five years.[11] Elsewhere, it is estimated that $1.6 billion in seafood enters Europe annually, and that approximately 50% of all seafood sold in Europe has illegal origins (see Table 4.1).[12]

While the practice of overfishing has drastically reduced some fish stocks, some fishing methods are by themselves extremely harmful to the environment. Restrictions on illegal dynamite fishing led to an increase in the practice, first documented in the Philippines in 1962. In the 1990s, dynamite fishing by local fishermen in Tanzania became commonplace—the use of both dynamite and hand grenades to catch reef fish in large quantities remains an ongoing problem there.[13] Driven by a demand for live fish to satisfy upscale restaurants and aquariums, the use of sodium cyanide has become common in Southeast Asia. (Over 500 million fish are traded annually for the aquarium trade, with most specimens coming from Asia and the Pacific to serve markets throughout Europe, the United States, and Asia; the trade itself is predominantly legal.) In seawater, cyanide breaks down into sodium and cyanide ions—this has the effect of rendering fish unconscious. Fishermen dive to coral reefs and spray the cyanide—at great

Table 4.1. Estimates of Annual Value of High-Seas IUU Catches

	Species Group	Annual Value ($million-est.)
Tunas and tuna-like fish	Bluefin	33
	Yellowfin, albacore, bigeye	548
	Chilean Jack Mackerel	45
Sharks	Sharks	192
Groundfish	Toothfish	36
	Cod high seas	220
	Redfish	30
	Roughy/alfonsino	32
Cephalopods	Squid	108
	Total	1,244

Source: Review of the Impacts of Illegal, Unreported and Unregulated Fishing on Developing Countries, Final Report (London: Marine Resources Assessment Group Ltd., 2005).

personal risk—and then collect the yield. The problem is that the combination of cyanide use and post-capture stress results in a 75% mortality rate within 48 hours of capture. This high mortality figure causes fishermen to respond with greater fishing pressure to meet the aquarium demand. In much of Southeast Asia, there has been a significant breakdown of coastal areas that were formally rich fishing grounds due in part to the use of sodium cyanide. Many fishing and diving areas already damaged by dynamiting have been ruined and lost through cyanide fishing around the Philippine islands, a principal source of fish for Western aquariums. Dendritic varieties of corals, which provide vital safe areas for young fish, are especially vulnerable to cyanide use.[14]

A broad range of marine life is negatively impacted by IUU fishing. Recently, tuna and other large pelagic have been targeted for their high market value (and have likewise been the object of conservation measures). In the International Commission for the Conservation of Atlantic Tuna (ICCAT) regulatory area, there are small amounts of IUU fishing for bluefin tuna—maybe 1% of the total tuna catch, while IUU fishing for bigeye tuna probably is about 5% of catch. Perhaps 5,000 to 10,000 tons of tuna is taken illegally in the Atlantic. But in the Indian Ocean, illegal catches account for close to 10% of fish caught, or about 130,000 tons annually. The IUU tuna catch in the Pacific has been estimated to be between 100,000 and 300,000 tons, with a value of $134–400 million. Other high-seas species commonly taken through IUU fishing include redfish in the North Atlantic, orange roughy around New Zealand and Australia, squid in the south-west Atlantic, and toothfish in the southern Atlantic.[15]

IUU fishing for salmon is an increasing problem, most notably around the Pacific Russian coast. Relative to the 1950s through the 1970s, poaching of salmon spawning grounds has increased—in Russia, there is practically no effective river protection. Around Kamchatka, commercial IUU fishing is endemic, an "economy within an economy" that employs tens of thousands of illicit workers.[16] During the fishing season, practically every settlement in Kamchatka turns into a salmon-poachers camp of organized groups and locals. Most IUU fishing for salmon in the region involves excessive quota violations—quotas may be exceeded by a factor of 10 in the lower reaches of some river basins, where visiting dealers purchase salmon roe on site. The annual illegal catch of salmon in waters around Kamchatka averaged 55,000 tons from 2002 to 2006, while around Sakhalin, the number is 80,000 tons. Much of the illicit salmon catch centers on the lucrative market for roe (salmon caviar, called "red gold," is 10 times more valuable than salmon flesh—in many cases, the fish itself is simply discarded). In just the period from May to October 2005, Russian authorities seized from illicit roe traffickers in the Federal Okrug territory 58.4 tons of salmon roe valued at 24 million rubles. Curiously, Russian drift net operations for "monitoring and scientific reasons" far exceed the numbers required for such purposes. Japanese and Russian IUU salmon fishers typically record their catch of lucrative sockeye, chinook, and coho salmon as less valuable pink and chum salmon—not surprisingly, the amount of sockeye imported to Japan exceeds the reported catch of the species. Other illicit methods involve registration of ships with forged documents in Russian ports, and the widespread practice of keeping several sets of documents aboard fishing vessels. Some companies use twin ships, where fishing is pursued by two vessels with the same names and registration/board numbers. While thousands of salmon violations are officially recorded by enforcement agencies, only a few dozen actual result in any kind of criminal sanction (typically small fines and suspended sentences—in 2006, only 0.7% of assessed damages were collected by Russian authorities from 4,433 violators).[17]

Sharks are especially vulnerable to overfishing because of their slow rate of growth, late age of maturity, and low fecundity. At present, of

591 shark species assessed globally, more than 20% are on the World Conservation Union's (IUCN) Red List of critically endangered, endangered, or vulnerable species.[18] But most shark fishing is not illegal—the illegal sector is mostly associated with the retention of fins (shark fin accounts for only 7% of the volume of shark trade, but is 40% of the *value*). IUU shark fishers most commonly target hammerheads and silky sharks. While illicit practices occur globally, hot spots appear to be off the coasts of Central and South America, the Western and Central Pacific, and the northern waters of Australia.[19]

Catch and trade of sharks has continued to trend generally upward reflecting strong demand for shark meat and fins—an increasing problem is the amount of shark by-catch as a consequence of longline fishing for tunas.[20] The global shark catch peaked in 2003 at 900,000 tons, declined to 750,000 tons in 2006, then increased to 780,000 tons in 2007. A recent study of the shark fin trade in Hong Kong estimated that the global shark catch is vastly unreported to the Food and Agricultural Organization of the United Nations—the true figure is probably between 1.1 million and 1.9 million tons annually. A study by the FAO estimated that 200,000 tons of sharks are also discarded for a variety of reasons, including unreported by-catch—so the true impact of commercial fishing on sharks is generally unknown.[21]

IUU fishing proliferates in the shallow seas off the coasts of Africa. These are especially rich ecosystems, and the poor countries in the region are unable to police their territorial waters that extend 200 miles out from the coasts. In a 2001 aerial survey of Guinea's territorial waters, 60% of the 2,313 vessels observed were committing offenses. Prior to 2000, there were 450 illegal incursions into Guinean waters, and in one village, five fishermen were killed when their small boat was destroyed by an industrial trawler. Elsewhere, recent aerial surveys of Sierra Leone and Guinea Bissau revealed illegal fishing rates of 29% and 23%, respectively.[22]

After the collapse of the central government in 1991, the ongoing civil war in Somalia led to rampant IUU fishing when ships descended from the European Union (EU), Japan, the Middle East, and the Far East to exploit waters devoid of any effective regulatory body. With the longest coastline in continental Africa and a rich marine ecosystem, as many as 700 unlicensed foreign vessels trawl Somali waters for tuna, lobster, shark, and deepwater shrimp. Another concern is that these foreign trawlers are responsible for the destruction of coral reefs and the unsustainable harvest and/or by-catch of dolphins, sea turtles, and dugongs. Some of the ships are heavily armed, and come into conflict with artisanal fishermen who are forced into restricted (and safer) waters—this results in overfishing, the destruction of traps and stationary nets, and an overall reduction in the vitality of local economies. In sum, the food security and livelihood of Somali coastal communities are compromised by IUU fishing.[23]

Foreign fleets from the EU, China, Japan, Russia, and Namibia exploit Angola's poorly patrolled coastal waters, where a serious depletion of fish stocks could be devastating (40% of the population is undernourished). Many vessels operate without a license, while others are licensed by corrupt Angolan authorities who permit foreign ships to overfish. In Angola, just three boats patrol a 1,650 km coastline and a 330,000-square-mile Economic Exclusion Zone (EEZ). Foreign trawlers routinely violate the 12-mile coastal area reserved for Angolan fishermen, resulting in violent conflicts with local fishermen. In one instance, a foreign vessel rammed a local boat, killing a fisherman, and in another case, two Angolan inspectors "disappeared" while on observation duties aboard large industrial trawlers. Angolan authorities have been physically attacked and have had their boats rammed and sunk by illegal trawlers. Only since 2004 have new air patrols and a longer range vessel made improvements in Angola, assessing fines, seizing some illegal boats, and establishing a monitoring system.[24]

ORGANIZED CRIME AND IUU FISHING

Especially in the case of high-value products, organized crime groups are involved in domestic poaching, most notably IUU fishing for abalone and sturgeon.[25] Russian syndicates, Chinese Triads, and other Asian gangs are involved in the illegal harvesting of abalone, believed to generate $80 million annually.[26] IUU fishing by organized crime syndicates in South Africa decimated abalone stocks and led to the closure of the fishery in 2008. Illegal fishing there is also linked to drug trafficking, money laundering, and racketeering. Economic losses in South Africa are significant—the closure of the fishery naturally decimated legitimate fishing operations, while criminals continue to smuggle abalone catches out of the country and export it from neighboring states, causing direct losses in export taxes.[27]

One of the more notorious illegal fishing enterprises involves the overfishing of Caspian Sea sturgeon. Caviar has a high value, is not bulky, is easily transported, and its origin is easily disguised—coupled with high demand and restricted legal supplies, these conditions provide enormous profits for illegal entrepreneurs. Caviar sells for $300 to $500 on the black market, but in the United States and Europe, it retails for 10 times that amount. Perhaps 10 times more sturgeon are caught illegally than legally, a practice that has led to a 90% drop in Caspian sturgeon catches.[28]

Russian criminal syndicates are estimated to earn $4 billion a year through the illegal exportation of some 2 million metric tons of seafood, mostly Caspian sturgeon and other seafood products to Europe, the United States, and Japan. In 1997, Japan imported about $1 billion in seafood from Russia, a figure six times the amount appearing in official Russian trade data.[29] Caviar smuggling bears all the hallmarks of large-scale organized crime, and is in fact a trade dominated by the Russian "mafiya"—smuggled in suitcases, beluga caviar retails for $4,450 per kilogram. The illicit profits are significant enough to breed violence, as officials attempting to halt caviar smuggling have been murdered. Leading up to 1997, some two dozen members of a Russian anti-poaching unit were murdered, including the wife and child of one official. In November 1996, 54 Russian border guards tasked with disrupting the illegal caviar trade were killed in an apartment building bombing.[30]

Although "traditional" organized crime groups are involved in illegal fishing, most notably in the case of sturgeon and abalone, the problem is nevertheless characterized by the participation of a wide range of transnational players who are not part of any particular organized crime group. In fact, illegal fishing is more often perpetrated by "legitimate" entities—opportunistic corporate trawlers and corrupt Third World government officials willing to decimate fish stocks for profit. Even in the case of the caviar trade, largely controlled by the Russia mafiya, companies specializing in caviar are heavily implicated in the importation of the product into Western countries. Typically, caviar is routed through states with weak regulations, such as the United Arab Emirates (UAE), where counterfeit labeling and fraudulent documents are applied to shipments. For example, the U.S. firm Caviar & Caviar printed counterfeit labels and applied them to caviar tins in the UAE, and then imported the cargo into the United States as legitimate Russian produce.[31]

THE DYNAMICS OF THE TRADE

Relatively high-profit margins associated with illegal fishing are complemented by numerous difficulties and shortcomings in national and international controls. Flags of Convenience (FOCs), insufficient monitoring, inadequate penalties, tax havens that provide IUU fishers with low taxes, confidential banking systems that allow for the formation of shell companies to launder illicit profits, corruption and bribery in national regulatory bodies, and a general lack of assistance from developed countries are all factors that inhibit implementation of national and international measures intended to combat illegal fishing.

First and foremost, IUU fishing thrives because of a lack of enforce-ment capacity. Patrol vessels are spread too thin in territorial waters (Economic Exclusion Zones, or EEZs), let alone the vastness of the open ocean. Monitoring and enforcement is very expensive, and many nations simply lack the resources to effectively control IUU fishing. In the European Union, monitoring fishing among its member states costs up to 300 million euros annually—about 5% of the value of all landings. In the case of the North Atlantic Fisheries Organization (NAFO), the cost of monitoring EU vessels in 2002 amounted to over 7% of produc-tion (fish brought to port). In any event, all monitoring is easily circum-vented, especially when bribery and corruption of government officials is involved. And, of course, individual nations must have comprehensive and enforceable domestic fishing regulations in place, which is not always the case in developing nations.[32]

The problem of IUU fishing is facilitated by the practice of fishing vessels that fly flags of convenience (FOCs), a practice described by the European Commissioner for Fisheries in 2004 as the "scourge of today's maritime world."[33] Under international law, the country whose flag a vessel flies is responsible for that ship. However, international law permits nation-states to issue FOCs to foreign vessels, typically for a relatively small fee (perhaps a few hundred or few thousand dollars)—but then they simply ignore offenses. The practice creates a huge loophole that allows IUU fishermen to circumvent management and conser-vation measures. The problem is that flag states either are not capable of or do not have the political will to exercise control over their vessels, a situation often exacerbated by the lack of a genuine link between the nation and the owner of the fishing vessel. Corrupt home governments may only realize a few million dollars annually by selling their flag to foreign vessels, but collect additional bribes in exchange for ignoring violations such as catch limits. Meanwhile, legitimate revenue from sales taxes that would normally go into public coffers is typically lost under FOC schemes—essentially, revenue rents to developing coastal states is lost to IUU fishers. Flag-of-convenience (also called open registry) countries notorious for abusing the FOC system and permit-ting the unsustainable exploitation of territorial waters include St. Vincent, the Grenadines, Panama, Belize, the Bahamas, Cyprus, the Netherland Antilles, Vanuatu, Georgia, Sierra Leone, Cambodia, and Honduras (to name a few).[34]

The system of leasing territorial waters is optimal for circum-venting fishing regulations. Typically, fishing vessels can register on the Internet. Some vessels re-flag several times a year to confuse sur-veillance (called flag-hopping). In many cases, FOC vessels are backed by shell companies, joint ventures, and hidden owners, making it difficult to identify and fine owners even when regulations are enforced. The confidentiality of some banking systems facilitates the system of hidden ownership. Not coincidentally, there is a significant overlap between FOC countries and those with strong bank secrecy laws—Antigua, Barbuda, the Bahamas, Barbados, Belize, Gibraltar, Liberia, the Marshall Islands, the Netherlands Antilles, Panama, St. Vincent, Tonga, Vanuatu, and the Grenadines are prominent examples. In 2001, Lloyd's Maritime Information Service listed over 1,300 vessels greater than 24 meters flying flags of convenience— about 10% of the world fishing fleet.[35] The following case study exemplifies the problem of FOC schemes:

The Case of the Condor: This 53m vessel, which was built in 1968, has undergone several previous names: *Arosa Cuarto* (1989), *Pesca-mex III* (1998), *Cisne Azul* (1999), *Viking* (2001), *Inca* (2003), and *Condor* (2005). In 2000, as *Cisne Azul*, she unloaded toothfish in Port Louis, Mauritius, over several months, and in April 2000 was refused entry to the Western Australian Port of Freemantle on suspicion of unregulated harvesting of toothfish. According to Lloyds Register Supplement of September 2000, *Cisne Azul* was owned by Arcosmar Fisheries Corp. But by the time the Lloyds 2003/04 Register was published, the vessel had changed names (becoming the *Viking*) and owners, to Jose Manuel Sangueiro Lopez, who is also known as the Vice-President of Alcimar SA. Alcimar SA share an office in Montevideo with another Uruguayan company Navalmar SA. It is alleged that Alcimar chartered the *Maya V*, which was arrested in Australian waters in January 2004 on suspicion of illegal fishing. Both Navalmar and Alcimar have links to the so-called Galician syndicate of illegal toothfish operators based in Spain, which also includes companies such as Viarsa Catera SA and Viarsa Fishing Co., based in Mauritius; a group of Panamanian companies, Pac Fish Inc.; Ocean King, based in Seattle, USA; and Thalasa SA, another Uruguayan-owned company based in Mauritius. In 2001/02, renamed *Viking* and flagged to the Seychelles, the vessel unloaded toothfish in Port Louis, Mauritius. The Seychelles reported to the Commission for the Conservation of Antarctic Marine Living Resources (CCAMLR) in April 2003 that they had cancelled the flag and fishing permits of the *Viking*. Nevertheless, the vessel was seen in Port Louis (named *Viking*), in June 2003 receiving provisions, fuel and bait, and then again in October 2003 and September 2004 (renamed *Inca*). In March 2005, the vessel was renamed *Condor* and flagged to Togo. She was seen fishing with five other vessels, flagged to Togo and Georgia, on the Banzare Bank, an area which had been closed to fishing by CCAMLR. An armed Australian vessel requested them to leave, but because the Flag States of these vessels are not members of CCAMLR, international law does not allow any additional action to be taken.[36]

FOC registration decreases the operating costs for ship owners, who can avoid regulations related to insurance, crew training, and safety equipment, and do not have to pay for licenses, observers, catch documentation systems, or vessel monitoring systems. In international waters, fishing regulations apply only to countries that are party to regional fisheries management organizations (RMFOs). This means that fishing trawlers need only re-flag to a state not party to a RMFO, and it is free to fish with no regard for regional agreements and conservation measures. In sum, FOC vessels are generally beyond the reach of international law.[37]

While international law permits states to deny access to their ports and specify conditions for landing, the uneven application of port controls internationally also facilitates IUU fishing on a large scale. Chile requires all foreign fishing vessels to comply with conservation and management regulations and to use a vessel-monitoring system. South Africa prohibits offloading of fish from vessels flying flags of convenience and makes use of blacklists. Similarly, other ports require fishing vessels to provide documents detailing the authorization to fish and the onboard catch. The problem arises because even though port inspections are standardized globally, the results of inspections are rarely passed on to RFMOs or flag states—this permits illegal fishers to simply move to other ports with less strict regulations. So, the key to enforcement is international cooperation, as unilaterally implemented port controls lead to a displacement effect. In fact, illegal fishing is largely dependent on the existence of so-called "ports of convenience" that provide essential services to IUU fleets—taking on supplies and fuel, and landing and shipping illegally caught fish that then enters the international market. Las Palmas de Gran Canaria (Spain) is a notorious example that serves the Atlantic IUU fleet—the port hosts a number of companies that operate illegal vessels, and serves as a gateway for illegal fish shipped to the EU market.[38]

Ports of convenience aside, enforcement of fishing regulations is hampered further because many IUU trawlers simply resupply at sea, rotating crews and staying at sea for months by transferring their illegal catches to transport ships (called reefers). Illegally caught fish are "laundered" by mixing them with legally caught fish already onboard reefers.[39]

THE IMPACTS OF IUU FISHING

The problem of IUU fishing is probably more severe than is known, due to underreporting of catches. Underreporting varies by species and by region, but unreported harvests probably range from 25% to over 100% of declared catches. A case study of 10 countries, mostly in Africa, found that total losses in IUU fishing during the study period was $372 million, or 23% of the declared value of the catch. Underreporting catches is in effect a violation of quotas, and complicates scientific tallies of fish stocks.[40]

The most direct economic impact of illegal fishing in territorial waters is the loss of the value of the catch to coastal nations. Loss of gross national product (GNP) is supplemented by the loss of revenue that could have been generated from legal fishing vessels, including landing fees, licensing fees, taxes, and other levies. Underreporting of catches by otherwise legal vessels also robs nations of revenue—underreporting is as high as 50% in Kenya and 75% in the shrimp fishery of Mozambique. A 2005 report estimated that the total value of IUU catches within national waters is $3 billion annually. Other macroeconomic effects include strain on national budgets and the loss of employment within fishing and fish-processing sectors.[41]

There are many secondary economic effects of IUU fishing. Illegal fishing reduces demand for fishing gear and boats. The downstream phases of the fishing industry are negatively impacted, including fish processing, packaging, marketing, and transport. Because most IUU catches do not appear to be landed within the country from whose waters the fish were taken, there are also losses in terms of port revenue derived from transshipment fees, port dues, vessel maintenance, and bunkering. Economic "multiplier effects" on investment and employment are the norm, and include budget pressure on national economies due to the costs of monitoring and controlling IUU fishing, a reduction in the value of catches for local fishing fleets, and health and safety risks when artisanal and industrial fleets conflict. In developing nations where fish is the major source of protein, the reduction of fish in local markets leads to malnutrition and food insecurity.[42] Additional human costs involve the exploitation of workers in low-wage countries, who are subjected to dangerous working conditions and physical abuse (see Table 4.2).[43]

In addition to the severe economic losses, illegal fishing produces unsustainable and negative impacts on target species and ecosystems,

Table 4.2. Economic Impacts of IUU Fishing

Parameter	Indicators	Impacts
Multiplier effects	Multiplier impacts on investment and employment	Direct and indirect multipliers linked to fishing reduced with the loss of potential activities through IUU fishing.
Expenditure on MCS	Annual expenditure on MCS linked to IUU fishing	IUU fishing places budget pressures on MCS/fisheries management.
Destruction of ecosystems	Reduction in catches and biodiversity of coastal areas	Loss of value from coastal areas, e.g., inshore prawn fishing areas, and from mangrove areas that might be damaged by IUU fishing. Reduction in income for coastal fishing communities.
Conflicts with local artisanal fleets	Incidences recorded of conflicts between IUU vessels and local fishers	Reduction in the value of catches for local fishing fleets. Possible increased health and safety risks because of conflicts between the artisanal and industrial fleets.
Conflicts with MCS officers and vessels	Armed resistance by IUU vessels to MCS enforcement	Spiraling loss of effectiveness of MCS activities. Costs of MCS escalate, and there is a loss in cost effectiveness of MCS.
Food security	Availability of fish for local consumption	The reduction in fish availability in local markets may reduce protein availability and national food security. This may increase the risk of malnutrition.
Contribution of fishing to GDP/GNP	Value added; value of landings	IUU fishing reduces the contribution of EEZ or high-seas fisheries to the national economy and leads to a loss of potential resource rent.
Employment	Employment in the fishing, fish processing and related sectors	IUU fishing reduces potential employment that local fleets may make to employment creation. This is likely to be a major factor only in respect to EEZ IUU fishing.

(Continued)

Table 4.2. (Continued)

Parameter	Indicators	Impacts
Export revenues	Annual export earnings	IUU fishing by reducing local landings and nonpayment of access dues will reduce actual and potential export earnings. This may also hinder surveillance activities.
Port revenues	Transshipment fees; port dues; vessel maintenance; bunkering	IUU fishing reduces the potential for local landings and value added.
Service revenues	License fees; exchequer revenue	IUU fishing reduces the resource, which in turn reduces other revenues from legitimate fishing services.

Source: *Review of the Impacts of Illegal, Unreported and Unregulated Fishing on Developing Countries, Final Report* (London: Marine Resources Assessment Group Ltd., 2005).

resulting in a reduction in biodiversity and ecosystem resilience. Damage to delicate mangrove areas and prawn dramatically reduces the potential to naturally restore local fish stocks. Aside from pressuring lucrative target species such as tuna and sharks, IUU fishing also depletes less lucrative stocks that are nevertheless critical food sources in marine ecosystems. The by-catch of longline vessels (legal and illegal trawlers) can be devastating to nontarget species, including turtles, sea mammals, sharks, killer whales, and seabirds. In southern oceans, illegal fishing kills 100,000 seabirds every year, including thousands of endangered albatrosses. The use of illegal fishing gear like gillnets can further damage marine systems—some IUU vessels even use explosives to keep whales away from fishing lines. The destruction of marine habitat can have especially far-reaching effects, as maerl, coral seagrass beds, and inshore shallow seas are settlement and nursery areas for other marine animals and juvenile fish.[44]

The social impact of IUU fishing on human communities is significant. Particularly in regions where fish is the major source of animal protein, illicit fishing contributes to food insecurity—this is true of many coastal states in Africa, including Sierra Leone, Angola, Somalia, Kenya, Guinea Bissau, Guinea Conakry, and Senegal. In areas where the continental shelf narrows toward the coastline, as in Liberia, industrial trawlers come very close to the shore, at times precipitating conflict with artisanal fishers. Conflicts between industrial vessels, legal and illegal, are also common in Africa's shrimp fisheries, as well as the inland fisheries of Senegal and Mauritania. Armed resistance to fishing surveillance and enforcement operations appears to be increasing in waters around Somalia and Mozambique, further aggravating the potential for human injury and death. A reduction in fish stocks in local waters can also reduce employment opportunities and a subsequent decrease in household incomes.[45]

RESPONDING TO IUU FISHING

The foundation to protecting fish species and stocks is appropriate and robust conservation measures. International measures to combat illegal fishing include a 2003 agreement to require the registration of any high-seas fishing vessel over 24 meters, but so far, only 25 countries have ratified the law. Of course, absent a global database, vessels can still rename and fly under a different flag with ease, so international cooperation remains the key to enforcement. In West Africa, a number of countries have formed the Sub-Regional Fisheries Commission to coordinate surveillance and information sharing as well as protocols on hot pursuit into territorial waters. In 2001, an international monitoring, control, and surveillance (MCS) network was established with the EU, the United States, and Japan joining (the network has 40 total members). The initiative has enjoyed some success, including the apprehension of several IUU vessels; however, the network is voluntary, informal, lacks resources, and has no dedicated, full-time staff.[46]

The United Nations drew up laws and regulations in the 1990s to combat IUU fishing, and in 2001, 110 nations endorsed the UN Food and Agricultural Organization's (FAO) International Plan of Action to Prevent, Deter, and Eliminate IUU fishing (IPOA-IUU). Under the IPOA, signatory states were expected to develop Plans of Action by 2004, but only six nations met the deadline—and a full third had not even begun to implement National Plans of Action years after passage. A 2003 meeting of the Round Table of Sustainable Development at the Organization for Economic Cooperation and Development (OECD) led to the establishment of a High Seas Task Force, with the aim of defining practical solutions to the problem of IUU fishing. In 2002 at the World Summit on Sustainable Development in Johannesburg, and at the G8 Summit in 2003, nations committed to the IPOA-IUU and pledged to eliminate subsidies that contribute to illicit fishing. Along with the IPOA-IUU, the Rome Declaration on IUU Fishing (2005) calls on developed nations to provide financial and technical assistance to poorer countries to develop MCS programs. Unfortunately, much of this amounts to rhetoric only, as many developing states still lack the resources to combat illicit fishing in their territorial waters.[47]

An international effort toward fisheries management involves the formation of RFMOs, created under international agreements to be responsible for the management of high-seas fisheries and fish that migrate through territorial waters of member states. Recently, RFMOs have developed whitelists and blacklists of vessels that are permitted or not permitted to fish in the waters of member states. Key to the success of RFMOs is their willingness and efficacy in sharing information and blacklists with other RFMOs and the international community at large to deter IUU vessels from simply shifting to ports of convenience. At present, many RFMOs have incomplete and incompatible lists and registers, and none provide a definitive source of information such as the registration history of individual vessels, port inspections, RFMO blacklisting, and a history of vessel ownership. RFMO vessel lists are derived from the flag state of the vessel in question—in the case of corrupt governing bodies, the information provided is probably dubious.[48] Only four RFMOs have remits that include all marine fish, which leaves vast regions of the high seas open to unregulated fishing on pelagic and demersal fish. There are additional high-seas fishery organizations that focus on individual species, such as the International Pacific Halibut Commission and the Convention on the Conservation and Management of Pollock Resources in the Central Bering Sea.[49]

Despite numerous national, regional, and international measures, significant gaps remain in fisheries conservation and regulatory enforcement. A good example involves a broad range of shark species—there is little effective management of shark stocks at the national or regional level, and few species are subject to international conservation initiatives. Where management plans are in place, they tend to be generic (as opposed to species-specific) and indirect, controlling finning of sharks rather than addressing catch and mortality. Ten shark species are listed in the CITES Appendices, with various Sawfish in Appendix I and Basking, Whale, and Great White Sharks listed under Appendix II (a few additional species are listed under the Convention on Migratory Species).[50] However, CITES Parties declined to extend protection to additional shark species at the 2010 meeting in Qatar, including the hammerhead, scalloped, and whitetip shark (hunted for their fins used in Chinese banquet soup).[51] In 2000, the FAO at the United Nations developed an International Plan of Action for the Conservation and Management of Sharks. But the plan is dependent on the voluntary implementation of individual national plans of action, so progress has been patchy. Also, because nations rarely report shark catches in terms of particular species (in 2007, only 20% of shark catch data reported to the FAO was recorded on a species basis), it is very difficult to draw definite conclusions about trends in shark fisheries globally.[52] Another measure to protect sharks involved the International Commission for the Conservation of Atlantic Tuna (ICCAT), which agreed to a recommendation that shark fins account for no more than 5% of the weight of shark on board ships—but this is difficult to enforce due to the processing methods and relative ease of concealment.[53]

Some evidence suggests that enforcement of fishing regulations is improving. In 2004, Australia outfitted a fishing patrol boat with deck-mounted .50 caliber machine guns[54]—not an overreaction, given the methods of some IUU fishing operators. (For example, after incursions by IUU vessels into Mozambique's Bazaruto Archipelago National Park, the Mozambican government sent military personnel into the area. An IUU longliner was apprehended, but not before volleys of gunfire were exchanged and a rocket-propelled grenade damaged the fly bridge of the offending ship.)[55] Aerial patrols are increasing and are more effective at surveillance, but must be supplemented with ocean-going patrols. For larger vessels, onboard observers are the most prag-matic way to ensure compliance. Verifying landings against logbook data are the key, but of course the Flag State must agree to permit an on-board observer—unlikely in the case of IUU fishing. (Ultimately, in nonterritorial waters, nations must be held accountable for fishing ships flagged to them.)[56] One high-tech enforcement measure involves the installation of satellite vessel monitoring systems (VMS), a relatively inexpensive way of maintaining surveillance of fishing vessels. Exact locations are regularly transmitted to a central monitoring center through Global Positioning System (GPS) technology. Unfortunately, commercial fishing trawlers have learned to manipulate the system by tampering with the onboard "blue box"—they transmit "false posi-tives," effectively concealing their location in order to fish out of season or in restricted waters.[57]

There are signs of international cooperation. Beginning in 2000, the United Kingdom's National Centre for Fisheries Surveillance and Protection (CNSP) trained three artisanal fishing communities in Guinea to monitor their own fishing grounds (industrial trawlers had long exploited the broad continental plateau along the West African coast, and in the village of Bongolon, five local fishermen were killed when their canoe was destroyed by a foreign commercial vessel). In Guinea, 30,000 small-scale fishers rely on the fish resource, which provides 51% of all animal protein consumed. The local fishers were equipped to patrol the 12-mile EEZ and advised to report infractions to the local CNSP surveillance station, which would then launch a patrol boat. Two years into the project, illegal incursions by industrial trawlers into the EEZ dropped by 60%. Moreover, the project was cost effective (the budget was only $20,000), as the patrols by local fishers allowed for targeted missions by the CNSP, whereas previously they had only made six or seven random patrols per month. By 2005, similar initiatives were planned for Gabon, the Congo, and Mauritania.[58]

An MCS program established by the European Union and the South African Development Community (SADC) provides training and technical assistance to agencies that monitor and control commercial fishing in Tanzania, South Africa, Namibia, Mozambique, and Angola. The group of West African coastal states comprising the Sub-Regional Fisheries Commission (SRFC) created a Surveillance Operations Coordinating Unit to develop joint air and sea patrols and protocols on hot pursuit in territorial waters.[59] Success in some areas has been notable: thanks to the robust MCS and observer system, coupled with a foreign fleet licensing scheme, IUU fishing in Namibian waters has been largely eliminated.[60]

In May 2004 a joint mission between the Angolan and Namibian ministries of fisheries and the SADC-EU Monitoring, Control and Surveillance Programme utilized a new patrol vessel to board 19 vessels and impound 6 for violations of SADC fisheries legislation. Air patrols were begun as well, revealing 29 Chinese vessels committing serious violations, including harvesting in areas reserved for artisanal fishing and fishing during closed seasons. A vessel monitoring system was set up on 70 trawlers, and a database of registered vessels developed.[61]

An exemplary prosecutorial success story involves Hout Bay Fishing Industries, a company with rights to the South African rock lobster fishery. The company admitted to knowingly overfishing rock lobster and hake between 1999 and 2001, and was convicted on 28 charges of violating the Marine Living Resources Act. A Hout director pleaded guilty to 301 charges of corruption for bribing fisheries inspectors, and in 2004, three defendants, including the former head of the company, were found guilty in a U.S. court of conspiracy and smuggling wildlife under the Lacey Act. The offenders were sentenced to one to four years in prison and fined $7.5 million, but a later ruling in U.S. District Court recommended that the South African government also receive $41 million in restitution.[62]

Demonstrable gains in fisheries management are possible. Unregulated fishing for tuna in the Atlantic declined from about 2,000 tons in 1999 to 500 tons in 2002 following trade-related sanctions and the implementation of a blacklist by the International Commission for the Conservation of Atlantic Tuna (ICCAT). Similar protections precipitated toothfish catch declines from 33,000 tons in 1997 to 2,600 tons in 2004.[63]

Additional steps toward reducing IUU fishing involve trade-related and catch-documentation schemes, measures that keep track of legally caught fish from the point of harvesting to the time it reaches the consumer at market. While such plans render the marketing of illegally caught fish more problematic, the traceability of fish is difficult (the identification of specific genetic markers is one possible solution). So far, only a few RMFOs have implemented trade-related catch-documentation plans. Still, there do exist nongovernmental traceability schemes like the Marine Stewardship Council certification, and even plans implemented by individual fishing outfits and retailers.[64]

Some RMFOs have enacted trade embargoes against certain countries whose vessels are known to be involved in IUU fishing—ICCAT banned or applied sanctions to the trade in bluefin and bigeye tuna from major flag-of-convenience countries such as Panama, Honduras, Belize, and Cambodia. An approach begun by the North Atlantic Fisheries Organization (NAFO) restricts landings of fish caught by nonmember vessels. Similarly, in 1999, the Commission for the Conservation of Antarctic Marine Living Resources (CCAMLR) implemented a catch-documentation scheme aimed at preventing illicit toothfish catches from entering markets in countries belonging to that RMFO. A "whitelist" of vessels authorized to fish by contracting parties was established, and only those ships on the list were permitted to sell certified toothfish. (While whitelists focus on catches and individual vessels, blacklists seek multilateral penalties against operators from blacklisted states; so while blacklists require strong monitoring to enforce, the burden of proof in whitelist systems falls on the vessel operator to establish that fish were taken in accordance with regulations.)[65]

A well-funded and strengthened MCS network, a global database of high-seas fishing vessels that includes information such as IUU prosecutions, a requirement that all high-seas vessels be fitted with centralized VMS systems, and restricting access to ports to only those vessels that demonstrate compliance with national, regional, and international fishing regulations remain integral to any successful IUU fishing reduction plan. Ports of convenience like Las Palmas, Spain should be closed or subjected to severe penalties, and the landing and transshipment of illicitly caught fish should be prohibited by all states. Transport and supply vessels that transship fish or provide services to IUU vessels should be restricted under national law. As always, the keys to enforcement are monitoring, control, and surveillance enhancement, control of at-sea transshipment, and strengthening of port restrictions. Closing loopholes in international law that allows nations to issue flags of convenience is probably the most important step that can be implemented to lessen IUU fishing.[66]

Market-based measures can play a key role: better labeling for fish and fish products so that consumers can choose to shun illegally harvested fish (dependent on robust trade and catch-documentation schemes), the application of tariffs on fish from countries that exploit the FOC system and are known to have vessels that engage in IUU fishing, easing the marketing of legally caught fish and species harvested by artisanal fishers, and even trade embargoes against certain nations that facilitate IUU fishing are all measures that can produce observable positive results.[67]

One economic reality that must be recognized and addressed is the fact that the overexploitation of some global fish stocks is caused in part by increases in the size of the world fishing fleet. This overcapacity of commercial fishing vessels, brought about largely by the subsidization of distant water fishing fleets, contributes to IUU fishing by artificially reducing the capital value of both old and new ships. While subsidization increases the profitability of fishing vessels, their artificially reduced value and the overall bulge in the number of ships means that a large number of extremely cheap vessels—costing hundreds of thousands of dollars, instead of millions or tens of millions—are available for purchase by those intent on illicit fishing. Likewise, the large number of aging vessels in the world fleet have nowhere to go except IUU operations, because their seaworthiness (or lack thereof) makes them unfit for operations in properly managed fisheries. Subsidies for buybacks and decommissioning should be avoided because this too will decrease the effective capital cost of ships, and thus have a negative impact on economic performance (not to mention the conservation of fish resources). Decommissioned ships should not be sold, but disposed of properly (ship-breaking entails an additional slew of environmental concerns—discussed in Chapter 2). To reduce the global fishing fleet—desirable from an economic *and* an environmental view—well-defined fishing rights should be implemented and effective resource and management programs deployed. Such approaches are likely to foster economic efficiency, reduce the perceived need for subsidies, and thereby decrease the bulge in fishing vessels that contributes to IUU fishing practices.[68]

Gaps in high-seas governance encourage IUU fishing. Since the extension of Economic Exclusionary Zones (territorial waters) is not politically feasible, a governance mechanism for the high seas is required—possibly a complete set of RFMOs, with more species-specific conventions. While most tuna and salmon are covered by RMFOs, few other species are, including almost all demersal fish that include orange roughy, alfonsino, sharks, and squid. As always, international cooperation and an *even* application of regulations are critical, as uneven rules and enforcement only lead to the displacement of illegal activity from strong-enforcement to low-enforcement regions. Finally, the problem of IUU fishing is at heart an issue of weak, incompetent, and/or corrupt governance within countries—political will not only to promote sustainable and responsible commercial fishing, but the will to foster governments characterized by integrity are key to successfully curtailing illicit fishing.[69]

NOTES

1. *Review of the Impacts of Illegal, Unreported and Unregulated Fishing on Developing Countries, Final Report* (London: Marine Resources Assessment Group Ltd., 2005); *Pirates and Profiteers: How Pirate Fishing Fleets are Robbing People and Oceans* (London: Environmental Justice Foundation, 2005).

2. *Pirates and Profiteers.*

3. *Pirates and Profiteers.*

4. *Study and Analysis of the Status of IUU Fishing in the SADC Region and an Estimate of the Economic, Social and Biological Impacts, Volume 2, Main Report* (London: Marine Resources Assessment Group Ltd. 2008).

5. *Pirates and Profiteers.*

6. *Study and Analysis of the Status of IUU Fishing.*

7. *Pirates and Profiteers.*

8. *Review of the Impacts.*

9. *Pirates and Profiteers.*

10. *Review of the Impacts.*

11. *Pirates and Profiteers.*

12. "International Crime Threat Assessment," *Trends in Organized Crime* 5, no. 4 (2000): 56–59.

13. *Study and Analysis of the Status of IUU Fishing.*

14. V. R. Pratt, "The Growing Threat of Cyanide Fishing in the Asia Pacific Region and the Emerging Strategies to Combat It," *Coastal Management in Tropical Asia* 5 (1996): 9–11; C. V. Barber and V. R. Pratt, "Poison and Profits: Cyanide Fishing in the Indo-Pacific," *Environment* 40 (1998): 5–34; John W. McManus, Rodolfo B. Reyes, and Cleto L. Nanola, "Effects of Some Destructive Fishing Practices on Coral Cover and Potential Rates of Recovery," *Environmental Management* 21, no. 1 (1997): 69–78.

15. Ibid.

16. Natalia Dronova and Vassily Spiridonov, *Illegal, Unreported, and Unregulated Pacific Salmon Fishing in Kamchatka* (WWF-Russia and TRAF-FIC Europe-Russia, 2008).

17. Ibid.

18. Mary Lack and Glenn Sant, *Illegal, Unreported, and Unregulated Shark Catch: A Review of Current Knowledge and Action* (TRAFFIC International, 2008); Mary Lack and Glenn Sant, *Trends in Global Shark Catch and Recent Developments in Management*, (TRAFFIC International, 2009).

19. Lack and Sant, *Illegal, Unreported, and Unregulated Shark Catch*; Lack and Sant, *Trends in Global Shark Catch*.

20. Lack and Sant, *Illegal, Unreported, and Unregulated Shark Catch*.

21. Lack and Sant, *Trends in Global Shark Catch*.

22. *Pirates and Profiteers*.

23. *Study and Analysis of the Status of IUU Fishing*; *Pirates and Profiteers*.

24. *Pirates and Profiteers*.

25. *Review of the Impacts*.

26. "International Crime Threat Assessment.".

27. *Study and Analysis of the State of IUU Fishing*.

28. C. J. Shivers, "Corruption Endangers a Treasure of the Caspian," *New York Times*, November 28, 2005, http://www.nytimes.com/2005/11/28/international/asia/28sturgeon.html (accessed September 22, 2010); Jessica Berry, "Armed Gangs Threaten World Caviar Stocks," *Sunday Telegraph*, July 18, 1999, http://www.highbeam.com/doc/1P2-19129915.html (accessed October 6, 2010).

29. "International Crime Threat Assessment," 2000; http://www.illegal-fishing.info.

30. *Trends in Organized Crime* 3, no. 2 (1997): 23–24; http://www.illegal-fishing.info.

31. Ibid.

32. *Pirates and Profiteers*.

33. *Pirates and Profiteers*, 10.

34. *Pirates and Profiteers*; *Review of the Impacts*.

35. *Pirates and Profiteers*.

36. *Pirates and Profiteers*, 11.

37. *Pirates and Profiteers*; *Review of the Impacts*.

38. *Pirates and Profiteers*.

39. Ibid.

40. *Review of the Impacts*.

41. Ibid.

42. Ibid.

43. *Pirates and Profiteers*.

44. *Review of the Impacts*.

45. Ibid.

46. *Pirates and Profiteers*.

47. Ibid.

48. Ibid.

49. *Review of the Impacts*.

50. Lack and Sant, *Trends in Global Shark Catch*.

51. "Wildlife Group Nixes Shark Protections," *Tribune Review*, March 24, 2010, A6.

52. Lack and Sant, *Trends in Global Shark Catch*.

53. Ibid.

54. *Pirates and Profiteers*.

55. *Review of the Impacts*.

56. *Pirates and Profiteers*.

57. Ibid.

58. Ibid.

59. Ibid.

60. *Review of the Impacts*; Ussif Rashid Sumaila, Morten D. Skogen, David Boyer, and Stein Ivansteinshamn, *Namibia's Fisheries: Ecological, Economic and Social Aspects* (Delft, the Netherlands: Eburon Academic Publishers, 2004).

61. *Pirates and Profiteers*.

62. *Review of the Impacts*.

63. Ibid.

64. Ibid.

65. Ibid.

66. *Pirates and Profiteers*.

67. Ibid.

68. *Review of the Impacts*; Colin W. Clark, Gordon R. Munro, and Ussif Rashid Sumalia, "Subsidies, Buybacks, and Sustainable Fisheries," *Journal of Environmental Economics and Management*, 50, no. 1 (2005): 47–58.

69. *Review of the Impacts*.

CHAPTER 5
Illegal Logging

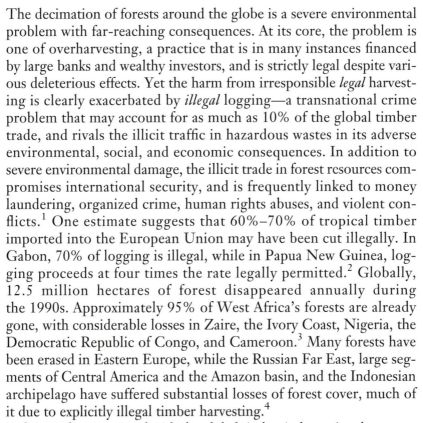

The decimation of forests around the globe is a severe environmental problem with far-reaching consequences. At its core, the problem is one of overharvesting, a practice that is in many instances financed by large banks and wealthy investors, and is strictly legal despite various deleterious effects. Yet the harm from irresponsible *legal* harvesting is clearly exacerbated by *illegal* logging—a transnational crime problem that may account for as much as 10% of the global timber trade, and rivals the illicit traffic in hazardous wastes in its adverse environmental, social, and economic consequences. In addition to severe environmental damage, the illicit trade in forest resources compromises international security, and is frequently linked to money laundering, organized crime, human rights abuses, and violent conflicts.[1] One estimate suggests that 60%–70% of tropical timber imported into the European Union may have been cut illegally. In Gabon, 70% of logging is illegal, while in Papua New Guinea, logging proceeds at four times the rate legally permitted.[2] Globally, 12.5 million hectares of forest disappeared annually during the 1990s. Approximately 95% of West Africa's forests are already gone, with considerable losses in Zaire, the Ivory Coast, Nigeria, the Democratic Republic of Congo, and Cameroon.[3] Many forests have been erased in Eastern Europe, while the Russian Far East, large segments of Central America and the Amazon basin, and the Indonesian archipelago have suffered substantial losses of forest cover, much of it due to explicitly illegal timber harvesting.[4]

Criminality associated with the global timber industry involves more than deforestation and harvesting trees in prohibited areas or beyond established quotas—illegal logging implicates "indigenous peoples' rights and public or private ownership rights; forest management regulations and other contractual agreements; transport and trade regulations; timber processing regulations, including the use of illegally harvested logs; and financial, accounting and tax regulations."[5] In sum, illicit timber harvesting and associated crimes contribute to deforestation and losses in biodiversity, drain government coffers and rob local communities of natural resources central to their livelihoods, distort forest product markets, and foster bad governance and violent conflicts.[6]

The trade in illegally sourced timber and wood products is fueled by demand in the United States, China, Japan, and Europe, and is maintained through poor governance, corruption, and violence. Because timber is an extremely bulky commodity that is nearly impossible to smuggle covertly and requires a specialized capacity to handle and process, the participation of large timber firms in the illicit sector of the logging industry is assured—as is the collusion of military, police, and government officials.[7] Various incentives drive the illicit trade: consumers who pay less for illegal timber products, poor communities proximate to forest lands that have few alternatives and depend on logging (legal and illegal) for their livelihood, and corrupt companies and public officials who reap enormous profits.[8] Of the $140 generated by every cubic meter of illegally harvested timber in the Russian Far East, nearly one-quarter is diverted to public officials, militia, and gangs in the form of bribes.[9] One estimate places the revenue generated by illegal logging at approximately $15 billion per year[10]—profits so great that the illicit revenue is used to help finance wars (called "conflict timber").[11]

The prodigious demand in China for wood products fuels a significant portion of the illicit timber trade throughout the world, including Southeast Asia, Russia, and Africa. The legitimate wood processing industry in coastal China is fed with illegal logs from Papua New Guinea, Burma, the Russian Far East, Indonesia, Tanzania, Gabon, Mozambique, Cambodia, Laos, and the Congo, while the furniture and wood products meet demand in Europe, Japan, and the United States. Since China instituted a domestic logging ban in 1998, imports of teak from Myanmar (Burma) total about 1.5 million tons (worth $350 million) every year.[12] Ikea, the Swedish home furnishing giant, is supplied in large part by Chinese manufacturers and Russian wood— but the company has only *two* foresters in China and *three* in Russia to ensure compliance with logging regulations. The World Bank estimates that half of all logging in the Russian Far East is illegal.[13] Elsewhere, Indonesia loses $1 billion per year in revenue and taxes as a result of illegal logging, and U.S. companies are robbed of $460 million a year as prices are depressed 7%–16% due to the infusion of illicit timber into the marketplace.[14]

The Indonesian archipelago has witnessed especially severe impacts from illegal logging. Approximately 65%–80% of Indonesia's timber production is illegal,[15] facilitated in large part by collusive arrangements among timber companies (most notably Asia Pulp and Paper, responsible for huge swaths of Indonesian deforestation), government officials, military personnel, and the police. Raw timber is exported from Indonesia and effectively laundered in Malaysia, Hong Kong, or Singapore, and from these locations, sawnwood, furniture, and plywood is reexported to global markets in Japan, Europe, China, and the United States. At the rate of cutting observed in 2005, nongovernmental organizations (NGOs) estimated that most of Indonesia's forests would be gone by 2015.[16]

The Tanjung Putting million-acre conservation park on Borneo Island, home to one of Indonesia's last surviving orangutan colonies, has been decimated by illegal logging and forest fires. About 10.5 million cubic feet of logs is stolen from the park annually. Again, public corruption is central to illicit logging in the archipelago, exemplified by the seizure in November 2001 of three Chinese-owned ships off the coast of Borneo with a cargo of $3 million worth of illegal logs—officials suspect that the timber belonged to companies linked to Abdul Rasyid, elected to Indonesia's Supreme Parliament in 1999.[17] But even when public officials do act with integrity and an environmental conscience, illicit practices are pervasive and difficult to stop. For example, when authorities attempted to halt the logging of the threatened merbau tree in 2002, shippers evaded the ban on merbau exports simply by forging documents and transporting the logs through Malaysia.[18] More recently, actions by timber giant Asia Pulp and Paper (APP) and associated companies demonstrate that Indonesian forestry law is violated with impunity: home to critically endangered Sumatran tigers, elephants, and orangutans, the Bukit Tigapuluh Forest Landscape in central Sumatra has been bisected by a logging road and areas of natural forest leveled in clear violation of regulatory law.[19]

Illegal logging is rampant in parts of Central and South America. Honduras and Nicaragua are among the poorest countries in the Western Hemisphere, and both have extensive forest cover. A number of factors ensure high levels of deforestation in these countries, not the least of which is institutionalized corruption in their forest sectors. In Honduras, one study found that clandestine production was estimated at 75%–85% of total hardwood output and 30%–50% of softwood production. In Nicaragua, clandestine production accounted for half of hardwood production and about 40%–45% of softwood production. Anecdotal evidence suggests that much of the remainder of timber in Nicaragua and Honduras is fraudulently "legalized" by mixing illegally harvested timber with legal logs.[20] In Peru, illegal loggers routinely invade the protected Alto Purus National Park to harvest a rare and valuable species of broadleaf mahogany. Government forestry agents tasked with inspecting lumber transported downstream on the Envira River are subject to violence and threats (their camp was burned to the ground in 2003).[21] Elsewhere in South America, Amazon deforestation was up 4% in the first half of 2008 after three years of declines—this despite government operations that seized illegal timber and levied millions in fines. A real concern is that restricting access to forest resources is not always popular with local populations attracted by the short-term benefits of illegal logging—in 2008, a mob of about 3,000 people angered over a crackdown on logging attacked a government office and environmental workers in the remote Amazon city of Paragominas.[22]

DIMENSIONS OF ILLEGAL LOGGING

The problem of illicit logging is exacerbated by the difficulty in distinguishing legal from illegal timber. Much of the production of forest products appears legal, but is merely "legalized" somewhere in the production chain. Such "legalized" production is, of course, fraudulent, and is a departure from national laws and standards that regulate timber extraction, processing, transport, and trade. Timber can be "legalized" at the stump simply by mixing in timber from an unauthorized region. Another good example of "legalized" production involved state policy in Honduras after the destruction of Hurricane Mitch in 1998—licenses were issued for "mahogany deadwood," a process that shortly evolved into fraudulent authorizations to cut standing mahogany. In Nicaragua, informal arrangements among state forestry agents, municipal officials, community leaders, and interest groups also facilitate so-called "legalization" of production—one method involves state officials who issue permits to cut larger volumes than physically available from an authorized region, thus informally encouraging harvesting in protected areas. In addition to "legalization" schemes, timber is produced globally through totally clandestine means—this sector of illegal logging escapes all documentation, forestry fees, and official statistics.[23]

An excellent case study of illicit timber production is the process as it has developed in parts of Central America. Illegal logging in Honduras and Nicaragua involves a broad network of actors, including community leaders, public officials, forest owners, sawyers, squatters, timber truckers, forestry professionals, and forestry industrialists. Institutionalized arrangements among these various players provides access to forest resources, up-front capital and equipment, transportation, processing and marketing, and formal and informal mechanisms that allow for the circumvention of the legal and fiscal system. Processors and exporters rely on timber traders and local contractors, who in turn advance capital and equipment to forest owners, timber associations, and individual sawyers. In Nicaragua, timber companies use intermediaries to buy community and/or noncommercial use permits in bulk, effectively limiting the participation of private forest owners and indigenous communities to giving permission for their land to be logged—essentially, local interests are largely divorced from the benefits of forest resource extraction. In Nicaragua, timber producers and forest owners receive only about 5%–10% of the timber's value, whether produced legally or illegally. As elsewhere, the illicit production of timber in Central America is largely dependent on the participation of government officials. "Institutional weaknesses" in the state forestry agencies in Nicaragua and Honduras translate into the fraudulent "legalization" of production by senior government officials. Another principal feature of the illegal timber trade in Honduras and Nicaragua is the trans-boundary movement of forest products. For example, data suggests a considerable under-declaration of hardwoods and softwoods out of Nicaragua and Honduras, especially to the Dominican Republic and the United States. Honduras also exports large quantities of illegal timber to Nicaragua, but then reimports it as "legal" timber.[24]

In addition to the loose network of "legitimate" and "quasi-legitimate" players that foster illegal logging, more "traditional" organized crime actors play a role. Especially in remote areas, the combination of unemployed youth, drugs, guns, and illicit timber contribute to the collapse of civil governance. For example, the remote Sico-Paulaya valley in Honduras is a major drug transshipment route and a refuge for urban gangsters. Ready cash from "traditional" organized crime activities like drug trafficking is invested in cutting and selling mahogany—some illegal logging operations are set up specifically to launder money derived from drugs and other organized crime enterprises. A raid on one unregistered sawmill in Honduras produced not only illegal timber, but stolen trucks and firearms. The participation of large-scale organized crime groups in illegal logging has been documented in the U.S. International Crime Threat Assessment (2000). The report states that well-organized criminal groups are involved in illegal logging, including organizations from "Africa, Asia, Latin America, China, Italy, Turkey, Afghanistan, Pakistan, and

Bosnia-Herzegovina." Russia has "timber mafias" in the Far East that purchase and deliver expensive timber to customers throughout Asia—these are unauthorized contracts for state-owned lumber.[25]

But the problem of illicit timber is not merely one of "organized criminals" and illegal entrepreneurs who operate on one side of a bright line between legality and criminality. Assessing the nature of the problem would not be complete without examining the role of export credit agencies (ECAs) that, at least indirectly, facilitate illegal logging. ECAs and investment insurance agencies are public or parastatal organizations that provide government-subsidized loans, guarantees, and risk insurance to companies looking to conduct business in countries where the business climate is too risky for conventional corporate financing. The idea is to boost the economic well-being of the subsidized corporation by facilitating its ability to win major export and construction contracts in foreign nations. During the 1990s, financing facilitated by ECAs amounted to $80–100 billion per year, about twice the world's overseas development assistance at that time. ECA involvement in illegal and unsustainable deforestation has been documented since the mid-1990s. The focus on "high-risk" and developing nations ensures that much of the capital investment deriving from ECAs goes to regions with weak institutional governance—nations characterized by high levels of public corruption with neither the capacity nor the inclination to effectively regulate industry and assess the sustainability of logging practices. While there is rarely direct ECA support for logging or timber trading, they do provide significant funds for the pulp-and-paper sector, thus indirectly contributing to illegal logging by boosting the demand for raw timber far beyond the local legal capacity to meet it. In fact, many of the pulp-and-paper operations in Indonesia were built with the assumption that illegal timber sources would be readily available (they are). This business environment not only facilitated illegal logging in Indonesia, but precipitated the unsustainable expansion of the pulp-and-paper giant Asia Pulp and Paper (APP). The company subsequently defaulted on the loans provided by the U.S., Japan, Canada, and six European nations, and a large chunk of Indonesia has been clear-cut to feed consumer markets in the developed world.[26]

Many policies of the World Bank indirectly promote illegal logging and contribute to significant deforestation—World Bank capital investment certainly generates incentives for overharvesting. In other cases, the link between illegal logging and World Bank funding/policy are more direct. A case in point is World Bank financing of palm oil plantations in Indonesia, a practice that contributes to the removal of 3.8 million hectares of forest annually. When the Indonesian economy collapsed in 1998, the International Monetary Fund (IMF) and the World Bank provided a relief package with conditions that lifted the ban on foreign investment in palm oil ventures—over the next decade, tropical forest cover about the size of Costa Rica was transformed into palm oil plantations, displacing forest communities and wrecking considerable environmental and social havoc. Beginning in April 2000, armed Indonesian police instituted raids, shootings, kidnappings, arrests, and torture to intimidate locals into giving up their land to a subsidiary of the Wilmar Group—the largest palm oil refiner and exporter in Indonesia, backed financially by the International Finance Corporation (IFC—the World Bank's private lending branch).[27]

In 2002, the World Bank adopted a new policy on forests with an objective of securing 200 million hectares of forest for regulated and certified sustained forest management. Although the World Bank plan was encouraged by and involved a partnership with the Worldwide Fund for Nature (WWF), many environmental NGOs and indigenous peoples' organizations condemned the policy because it did not apply to arms of the bank such as the IFC and the Multilateral Investment Guarantee Agency (MIGA)—and it removed the proscription on funding for logging in tropical forests. The new policy also relies on uncertain certification standards and permits World Bank funding of forest clearance for plantations. Concern over the 2002 World Bank Forests Strategy appear to be well founded, as many Bank promises, including certification standards, have not been implemented—in fact, in cases where wood certification standards are required, forest sector loan and grant recipients are notably lacking. Moreover, the World Bank's "reforms" have led to demonstrable negative consequences for forests and indigenous peoples. "Community Forestry" projects in India meant to lessen poverty have instead exploited indigenous populations and forcibly evicted locals from their lands, typically with little or no compensation. That World Bank loans to governments come with nefarious strings has been well documented—for example, the Indian government and the Andhra Pradesh Forest Department (APFD) coerce local communities into providing commercial access to private plantation, pulp, and mining concerns. Promises of profit sharing typically involve the dubious provision of low-wage forestry jobs, while "microplans" for forest management and community development are farcical (adopted without prior agreement or input from locals, who are denied access to the relevant documents). In Andhra Pradesh, requests by villagers to plant a mix of native species and fruit trees was rejected by the Forestry Department in favor of eucalyptus, favored for its commercial value as pulp wood.[28]

Elsewhere, the World Bank and the United Nations Food and Agricultural Organization are planning on carving up much of the Congo basin into zones that include vast areas set aside for logging concessions. In Cambodia, initiatives to eliminate forestry-related corruption have floundered due to lack of Bank support. Meanwhile, dubious logging projects in the Amazon basin have been funded by the IFC and MIGA. When the IFC loaned $50 million in 2004 to the Brazilian company Aracruz Celulose, bank officials ostensibly did not consider the position of indigenous Tupinikim and Guarani, who have been fighting Aracruz and the Brazilian government for decades over lands forcibly taken from them. Despite the fact that soybean farming and cattle ranching has cleared 80 million hectares of Amazonian rainforest and caused massive environmental degradation and violent conflict between local tribes and state-backed corporate interests, the IFC loaned $30 million to the largest soya agribusiness in the country in 2002 and 2004.[29]

ENVIRONMENTAL, ECONOMIC, AND SOCIAL IMPACTS OF ILLICIT LOGGING

The environmental consequences of illegal logging are enormous. Harvesting trees without considering environmental consequences has decimated some tree species, while the loss of habitats has likewise driven some animal species to the brink of extinction. Merbau, teak, and mahogany have all witnessed serious declines, and ecosystem disruption has endangered tigers in Sumatra, orangutans on Borneo, and tigers, red pandas, and leopards in Burma.[30]

Deforestation from illegal logging also contributes to soil erosion, a problem that alters hydrological systems and reduces hydro-biological potential. The loss of trees leaves a thin remaining layer of topsoil, which amplifies water runoff—consequently, the soil loses nutrients, and silt builds up in rivers and estuaries, damaging delicate ecosystems including mangroves and coral reefs. Illegal logging also contributes to the loss of rare plant species, including those with valuable pharmacological properties. Losing tropical flora reduces the ability to breed genetic defenses back into food source plants that have become susceptible to disease.[31]

In addition to severe environmental harm, illegal logging produces significant economic losses. In Honduras and Nicaragua, the monetary damage goes far beyond the estimated $2–12 million in direct annual fiscal losses. Accounting for declining timber production in line with a constant deforestation rate, the "net present value" losses for Honduras have been estimated to be $58–91 million, and $16–30 million for Nicaragua. Moreover, such figures are clear underestimates, as they exclude additional sales tax and income that would result from higher timber prices (an economic reality if illegal timber production did not exist or was reduced as a proportion of market share).[32]

Globally, environmental and economic losses due to illegal logging are linked. Opportunity costs (expenditures that would not be necessary in the absence of illegal logging) associated with monitoring and law enforcement siphon funds that could otherwise be used to foster sustainable forest management. Moreover, the corruption associated with illegal logging directs private investment into rent-seeking investments, and away from activities with high social, environmental, and economic benefits (like sustainable forest management practices). Revenue from industrial-scale illegal logging is typically expatriated, resulting in additional national losses due to the economic "multiplier effect." As macroeconomic analyses consistently demonstrate, there is a strong correlation between weak governance and social wellness indicators such as per capita income and infant mortality. In short, corruption and impaired governance deriving from the illicit timber trade can be seen as contributing directly to broader societal problems. Corruption and patronage among state forestry officials impairs governance, undermines environmental monitoring, and distorts the roles of key actors in the timber production chain. Under circumstances in which large-scale timber merchants purchase the acquiescence of public officials in illegal resource extraction, community organizations and local interests are easily bought out through a combination of bribes, credit, and intimidation.[33]

Local poor and indigenous populations are most negatively impacted, as they disproportionately feel the loss of state revenues, nonmarket benefits, and the breakdown of justice in rural areas. Benefits to local populations tend to be temporary. For example, the production of a huge quantity of illicit timber in Honduras' Sico-Paulaya valley in 2000–2001 earned the local population about $1.2 million, but the overall effects were largely insignificant, even negative. Powerful community members seized the majority of the money, the intense economic activity for two years was followed by a dramatic decline, and several local chain saw operators and other businesses actually ended the period in debt.[34]

Negative impacts on impoverished local populations are not merely economic, but cultural and social as well. Illegal logging has divided and alienated entire communities. In Honduras, rival chainsaw gangs have emerged, and in Nicaragua, indigenous leaders have been corrupted and traditional institutions eroded.[35] Logging in Papua New Guinea (PNG) is dominated by a small number of Malaysian companies, the most prominent being Rimbanan Hijau—a name "synonymous with political corruption, police racketeering and the brutal repression of workers."[36] The actions of the company routinely destroy food sources, water supplies, and the cultural property of indigenous communities, all the while providing a climate for corruption, violence, and arms smuggling. Human rights abuses in PNG are extreme and are facilitated by the PNG government, which collusively arranges the theft of forest lands from local landowners. Those who stand up to government-backed loggers are subject to arbitrary detentions and physical abuse by police. Working conditions for the local labor force is akin to slavery workers who die on the job are often buried on site so the company does not incur the expense of transporting the bodies home. On a larger scale, regional security is at stake, as there remains significant cross-border traffic in smuggled timber, guns, and people. Meanwhile, Malaysian companies conceal the theft of lucrative merbau logs from West Papua by labeling them "PNG" despite their Indonesian origin. Cargo vessels fly false flags, port authorities take bribes, and Papua New Guinea Defense Force soldiers provide security for the illicit timber shipments.[37]

An excellent case study of the negative impacts of illegal logging involves the situation in Peru's Alto Purus National Park, a vast stretch of forest wilderness in the Amazon watershed that holds large stands of rare and exceptionally valuable broadleaf mahogany trees. The region also harbors nomadic hunters and gatherers: hundreds of indigenous people who live in voluntary isolation, they are among the last "uncontacted" people on the planet, hunting with bows and arrows and residing in temporary shelters constructed from palm fronds. Since 2004, the core of the region—an area about the size of Costa Rica—has been protected by the Peruvian government as well as international laws that protect various species (in 2002, mahogany was listed in Appendix II of CITES—see Appendix B). Nevertheless, since Brazil ceased exporting mahogany in 2001, Peru has become the world's largest exporter—so illegal logging continues, and threatens the flora, fauna, and uncontacted peoples inside and along the borders of Alto Purus. By 2004, Peru's mahogany range had decreased by 50%, and in 2006, flyovers and expeditions up six rivers uncovered numerous active logging camps in prohibited areas. Policing the Alto Purus can be dangerous: forestry engineers tasked with checking the legality of timber flowing downstream are threatened with violence (one post was destroyed by arson in 2003), while violent incidents between uncontacted peoples and loggers have become more common, with deaths on both sides.[38]

In and around Peru's Alto Purus park, local communities are systematically exploited. Loggers have established a system where they offer marked-up equipment and commodities (in advance) to communities in exchange for permission to cut trees, but locals lack the knowledge and negotiating skills to engineer a fair exchange. So, *illicit* timber aside, the opportunity for legal, responsible and sustainable harvesting of mahogany resources is lost to impoverished locals while illegal loggers and their corrupt patrons reap the proceeds (somewhere along the supply line from harvest to the port in Lima, exporters obtain permits for wood that is not in compliance with CITES regulations, a fact that clearly suggests official corruption). Some villages are so poor they have no schools, wells for clean drinking water, or latrines—yet in the nearby forest some mahogany trees felled by loggers (each worth tens of thousands of dollars on the international market) are left to rot because of small holes in the trunk, which reduce their commercial value. Of course, it must be recognized that, unscrupulous logging companies and corrupt public officials aside, the entirety of the Peruvian illicit mahogany industry (and the illicit timber trade in general) is driven by demand for high-quality furniture in developed nations.[39]

CONFLICT TIMBER

While its bulk does not make it an ideal commodity for financing arms purchases and war, timber is nonetheless commonly used in this capacity. So-called "conflict timber" is conflict financed or sustained through the harvest or sale of timber, or conflict emerging from competition over timber and other forest resources. Illegal logging naturally arises in heavily forested regions characterized by conflict. Moreover, there is a direct link between conflict timber and poor, inequitable systems of governance. Failed and failing states where the government is incapable of systematically making and applying rules that citizens will accept as legitimate enhances illegal timber production and the subsequent growth in illicit revenue used to finance conflicts.[40] Another factor that produces conflict timber is the incidence of ambiguous land tenure claims in Third World forested areas. In the case of weak or failing nation-states, governments unwilling or unable to adjudicate competing land claims breed conflict, while vast tracts of forest provide the means to finance it. In such cases, the stronger of the two parties, typically an agent of the state, wins the land battle—and then trees are cut with little or no regard for negative environmental, economic, and social impacts.[41]

Using timber to finance conflicts—and associated illegal logging— has been observed around the world, including in Afghanistan, Burma, Cambodia, Indonesia, Liberia, Vietnam, Nepal, and the Philippines. On the El Salvador–Honduran border, the illegal timber trade is worsened by old territorial disputes. Timber conflicts there have involved local authorities and well-armed local groups. In Vietnam, the government is settling ethnic Vietnamese in rural highlands to control indigenous hill groups suspected of seeking independence—the process has brought the two groups into conflict over forest resources. The government has encouraged the ethnic Vietnamese to log large areas of forested land to plant coffee, impoverishing the forest-dependent *montagnard* groups and sustaining that conflict.[42]

The governments of Liberia and Burma have supported illegal logging activities that generated revenue used to finance prolonged conflicts, resulting in the decimation of forests and the death and displacement of large numbers of people. In Liberia, President Charles Taylor has used conflict commodities like diamonds as well as timber to finance military operations domestically and across the Mano River states (Sierra Leone and Guinea). Taylor authorized a small number of firms to harvest timber for cash; alternately, the logs were bartered to Chinese traders directly in exchange for munitions. By mid-2003, Guinea alone was thought to harbor some 200,000 Liberian refugees displaced by Taylor's timber- and diamond-financed wars. Sierra Leone, Guinea, Gabon, and the Democratic Republic of Congo all have conflict timber interactions—either timber-financed wars or competitive strife over timber resources.[43]

In Burma, the military junta has sought to neutralize hill tribes resistant to central control by authorizing Thai logging firms to harvest teak along the Thai-Burmese border. The Burmese regime also negotiated logging concessions with China for the purpose of creating logging roads into highland tribal areas—thus providing the military access to the remote tribes. The hill tribes have countered through logging of their own, or by authorizing Thai firms to harvest trees in exchange for licensing fees used to finance their resistance. In this conflict, as elsewhere, environmental consequences and sustainability are nonissues.

In Cambodia, forest cover decreased from 75% of land mass in the early 1970s to less than 35% by the mid-1990s. Most of the loss was due to illegal logging, sanctioned by the Royal Cambodian Armed Forces and the Khmer Rouge in order to generate arms purchases. In a typical month in 1992, the illicit timber trade generated $10 million for the Khmer Rouge. In 1995, Cambodia's two prime ministers secretly granted logging concessions to 30 (mostly foreign) companies to harvest some 6.3 million hectares—an area three times the size that could support legal commercial logging. The $117 million generated bypassed the national budget and was used instead to help finance the reelection campaigns of the prime ministers. The July 1997 coup in Cambodia was provoked in part by illegal timber revenues that had helped pay for imported weapons.[44]

While government opponents and insurgencies set up partnerships with illegal loggers to finance their rebellions (as in Myanmar/Burma), national governments are almost always complicit in conflict timber activities. In fact, in most cases, government military units and state-backed logging companies foster and benefit from conflict timber production, as they have the resources and capital in the first place to harvest and transport such a bulky commodity. Moreover, loose financial oversight creates incentives for powerful actors such as the police, military officers, and politicians to engage in illegal timber production for the purpose of financing military operations. In a variety of case studies, researchers found that government and government-backed actors had unregulated access to "private" banks, money transfer shops, and other channels (including bulk money transfers across porous borders) that guaranteed the movement of money out of the source timber countries.[45]

In sum, illegal logging and conflict timber divert resources from legitimate government and indigenous populations, and undermine legal timber operations that promote forest sustainability. Whole communities and traditional ways of life are displaced or destroyed, and, as always, extensive deforestation contributes to habitat destruction, loss of biodiversity, soil erosion, siltation of rivers, and damage to mangroves and coastal reefs.

LAWS AND ENFORCEMENT

Reducing or halting illegal logging and the associated trade in illegally sourced wood products is confounded by a number of variables, not the least of which is the significant profits to be earned. A number of legal and institutional factors actually inhibit legal timber production. First, there is the perception that laws governing forests are illegitimate and transitory (assigning harvesting rights to third parties, for example). Other significant impediments to legal logging is unfair competition from state-backed illicit operators, and unclear and complex regulations—a formula for crime that places legitimate timber concerns at an even greater disadvantage relative to illicit loggers. A case in point is the COATLAHL timber cooperative in Honduras, driven to bankruptcy (and illegal logging) by relatively high production and transaction costs due to overly complex state regulations and the theft of timber by armed gangs. Yet another factor contributing to illegal logging is the existence of overlapping and conflicting

governmental responsibilities, a condition that produces legal uncer-
tainties as well as ample opportunity for corporate interests and corrupt public officials to circumvent the law. Finally, lax enforcement
and weak penalties virtually ensure that illegal logging remains a
highly lucrative endeavor.

In just the last decade, encouraging signs have developed indicating
that the international community is viewing the problem of illegal logging more seriously. One important development has been the implementation of high-level ministerial forest law enforcement and
governance (FLEG) processes begun in Asia, which led to the Bali Declaration and FLEG action plans in East Asia (2001), Africa (2003), and
Europe and North Asia (2005). In 2003, the European Union adopted
the Forest Law Enforcement, Governance and Trade (FLEGT) Action
Plan with the aim of ending all illegal timber imports into Europe.[46]
The World Bank, while bankrolling questionable logging practices
(detailed above), has nevertheless played an important part in fostering
these regional plans by aiding governments in establishing independent
log tracking and forest monitoring tools, as well as illegal logging action
plans. As a consequence of World Bank influence, ministerial declarations have been endorsed by consumer and producer nations, civil society, and the private sector. A recent World Bank report suggested
employing asset forfeiture and anti-money-laundering laws to thwart
timber crimes and related corruption. Noting the dearth of successful
timber law prosecutions globally, the World Bank also organized a
meeting of experts in 2006 to explore law enforcement issues and possibilities—discussions focused on statute-of-limitations issues, corporate liability, proof of intent, the use of circumstantial evidence, and
the evidentiary use of photographs and computer analysis to aid in log
tracking and certification.[47]

In the United States, amendments to the Lacey Act passed in
May 2008 make that law the world's first to comprehensively ban the
import, export, or trade in illegally sourced plant products, including
timber and wood products.[48] In February 2009, the European
Parliament (EP) voted to impose stiff penalties on companies dealing
in illegal timber. Fines would be at least five times the value of the timber in question, and authorities would also have the power to seize the
logs and shut down the most egregious violators. (At the time of this
writing, the measure was yet to be voted on at the EP's plenary session, and only then would a "yes" vote send it on for review—and
likely alteration—by the European Union's agriculture ministers.)[49]

There has been considerable progress in eliminating the illegal
timber trade in Bolivia and Ecuador. In particular, Bolivia may be
exemplary in its efforts to maximize governance in its forest sector

by developing straightforward and accessible regulations—and this while the nation ranks high on Transparency International's corruption index. Obviously, while reducing corruption and promoting responsible governance is desirable, the experience in Bolivia suggests that it is possible to make progress toward sound forest management even when nations are generally plagued by exploitative public officials.[50]

Movement in a positive direction, while encouraging, has fallen far short of palliation. There is, in fact, considerable disagreement about how to address the problem of illicit timber, including how best to define terms like "illegal logging" versus "unsustainable logging."[51] Critics of FLEG processes, most notably the broad array of environmental NGOs, argue that framing the problem of unsustainable logging in law enforcement and governance terms merely works to provide a veneer of legitimacy to corporate loggers and complicit government officials who engage in irresponsible behavior—their point being that defining a behavior as legal does not automatically translate into environmentally, socially, and economically sound practices.[52] There is also considerable disagreement over the utility of voluntary partnership agreements (VPAs)[53] versus outright trade bans. An advantage of import bans is that such laws place the burden of proof on the importer to establish that the timber or wood product was legally harvested and procured.[54] On the other hand, framing illegal logging as merely a consumer problem by banning imports outright may ignore the "carrot" option of carrot-and-stick approaches, while simultaneously fueling a black-market timber trade. Of course, there are critics of VPAs as well—the EU FLEGT action plan *mandates* nothing, while the agreement in no way prevents *third-party* countries from importing illicitly sourced wood and wood products.[55] VPAs do not require an independent audit, generally cover only a few basic products (like sawnwood, roundwood, and plywood), and are confined to producer countries.[56] Some parties contend that without comprehensive legislation that prohibits illegal wood imports and promotes sustainable forest management globally, VPAs will simply fail.[57]

Clearly, the challenge of sustainable forestry management involves a careful balancing of competing interests, and a more precise definition of "sustainability." Curbing illegal logging and promoting sustainable timber harvesting must involve the right balance of incentives and disincentives, and employ both supply-side and demand-side strategies. Strict regulations and penalties for noncompliance have their place as a deterrent, but must be applied with discretion and tempered with incentives that encourage timber companies to act responsibly. Impoverished communities that depend on forest resources should not have their livelihoods criminalized by unfair or archaic laws, nor should small businesses be driven out of the market by regulations they cannot afford.[58]

Credible certification and labeling systems are critical to an effective demand/consumer-oriented strategy to combat illicit logging. The effective tracking, labeling, and certification of forest products from harvest to end use is essential, as responsible importers and consumers can then make sound judgments about the wood products they purchase.[59] The adoption of so-called "green procurement" policies in import countries is a critical component of demand side strategies as well.[60] A sound example of wood certification is the Tropical Forest Trust (TFT), whose criterion for legality is that the wood can be traced back to legal logging operations. Verification involves the monitoring of wood control systems (WCS) at factories tied to TFT-supported forest projects. TFT member companies operate the WCS, and TFT staff monitor the supply chain at every point at which illegal timber could be introduced—from the standing tree, through the forest, on to the factory, and through all stages of the manufacturing process. Because it is difficult to verify the origin of logs bought from timber traders on the open market, TFT staff link factories to forest managers and aid them in buying logs directly from the forest—this effectively shortens the supply chain and renders the process more transparent and easy to monitor. WCS systems were first used in Vietnamese garden furniture stores. While the furniture industry doubted that large factories could assume control of log supply chains, within two years of implementation, all TFT member companies buying furniture in Vietnam could trace the product back to known legal forest outfits. The success in Vietnam led to the introduction of WCS in Indonesia and at Chinese plymills. Significantly, the success of WCS is dependent on certification by *independent* third-party auditors.[61] Additional initiatives in the area of certification and verification include the use of "TracElite" satellite systems and barcode log-tracking technology.[62]

Advances in the certification and verification of timber and forest products is not the only encouraging development, as models for successful forest management do exist. Moreover, the natural tensions between economic activity/natural resource extraction and conservation/environmental concerns need not produce outcomes in which there is a "loser." As the following case study illustrates, "win-win" scenarios are practicable and within reach.

> The Brazilian state of Acre ... has been a major leader in the search for environmental and social policy alternatives. Prior to 1999, the state's forest sector focused on exporting raw materials

for use in other states' industries. This left Acre with an under-performing forest sector that was operating almost totally in illegality. This illegality was made possible by a lack of control by the state government, the availability of financial incentives for cattle ranching, and the economic crisis of the extractive sector that was driving local communities to sell their forests to farmers and loggers . . .

The first period of the "forest government" [an administrative unit put in place by Acre governor Jorge Viana] focused on the revitalization of the extractive economy through the subsidization of natural latex production, the reorganization of forest productive cooperatives and the construction of industrial plants for improving latex and Brazilian Chestnut industrialization . . . The government also began the economic-ecological zoning of the state, which established the basis for land-use planning. The government also invested in the state environmental body, firing corrupted employees and hiring newly trained professionals. It was during this period [1999–2002] that the Federal government gave the state control and licensing authority for forest-clearing activities.

Since 2003, the "forest government" has intensified its forest policies, with massive investments through its Sustainable Development Program, financed by the Inter-American Development Bank and the Brazilian Development Bank. These investments have generated an impressive set of outputs, including the creation of approximately two million hectares of Conservation Units, the declaration of three State Forests, the creation of a community-based forest management programme, fiscal incentives for companies that invest in sustainable forest management, and support for forest industries that produce high value-added products. Acre has now become the second state in the Amazon Region to obtain complete decentralization of its forest administration, enabling it to expand its control to private forest lands.

The results seen in the state's forests are equally impressive. Acre has maintained more than 90 percent of its original forest surface and kept its average deforestation rate down at around 0.3 percent per year. *This is alongside an economic growth rate of 5.3 percent—twice as high as the national average* [emphasis added].[63]

The vast range of environmental nongovernmental organizations (NGOs) will continue to play a critical part in the fight against illegal logging. A prime example is the World Conservation Union (IUCN), an entity at the forefront of practicable and sustainable forest management. Acknowledging the complexities of the global timber trade and the disparate interests involved—not the least of which are countless poor communities dependent on timber resources for their livelihoods—the IUCN recommends a "tripartite" approach in which multiple stakeholders are brought to the table and the varied interests represented.[64] Well-funded NGOs can also bring to bear considerable pressure on private logging corporations and governments complicit in illegal and unsustainable practices. For example, dozens of NGOs led by Greenpeace precipitated boycotts of Asia Pulp and Paper (APP) products when the company began to illicitly clear the Simao forest in China's biologically diverse Yunnan Province. A lawsuit filed by APP against the Zhejiang Hotel Association, which backed the boycott, was dropped due to domestic and international NGO pressure.[65] The International Tropical Timber Organization (ITTO) provides the only tropical timber commodity agreement between producer and consumer nations—illegal logging remains high on the ITTO's action agenda.[66]

The Convention on International Trade in Endangered Species of Wild Fauna and Flora (CITES), the principal international agreement among governments to combat wildlife trafficking, can also be used to combat illegal logging. A number of tree species are listed in the CITES Appendices, including Alerce, the Monkey puzzle tree, Cuban or Spanish mahogany, Afrormosia, and ramin. Agarwood, a non-timber species valued for its fragrance, is also regulated. CITES has in place a monitoring system for inspection of listed species at both import and export sites, and has the authority to regulate tree species across all 167 of its Parties.[67]

Finally, in combating illegal logging, it must be recognized that a one-size-fits-all approach will not generate desired outcomes, and would likely produce unintended consequences. Market-based analyses reveal that the causes, structures, and methods of illegal logging differ by region and are dependent on economic, political, and ecological variables. There is in fact a systematic variation in illicit timber activity across regions. For example, while some level of governmental involvement is typical, participation ranges from nationwide wholesale corruption to petty corruption at the local level to mere indifference and/or the inability to monitor and regulate the trade. In some cases, illegal logging may be defined best as forms of white-collar or organized crime; and yet in other circumstances, loggers, processors, and timber transporters are otherwise law-abiding, but are obliged to operate outside the law due to ambiguous and contradictory regulations and ill-defined property rights.[68]

Disparate national laws that may vary depending on the specific tree species, disagreements over approach, vague and changeable definitions of legality, the complexities of certification and verification schemes, and the liberalization of international trade all confound the problem of illegal logging. Models for success do exist, but global progress in halting the illicit timber trade will certainly require unprecedented international cooperation. At the very heart of the problem is the overconsumption of timber and wood products and the collusive behaviors of private entities and governmental actors who apply investment capital in an unsustainable and irresponsible manner. Real change will require not only behavioral changes associated with consumption, but also that private and public elites more fully develop a social, economic, and environmental conscience.

NOTES

1. James Hewitt, *Failing the Forests: Europe's Illegal Timber Trade*, (WWF-UK, 2005); Rob Glastra, *Cut and Run: Illegal Logging and Timber Trade in the Tropics* (Ottawa, Canada: IDRC Books, 1999).

2. Duncan Brack, "Illegal Logging: Briefing Paper." Chatham House: Energy Environment and Development Programme, 2007, http://www.euflegt.efi.int/uploads/16LaceyActbp0702.pdf (accessed November 1, 2010)

3. *Illegal Logging, Governance, and Trade: 2005 Joint NGO Conference* (FERN/Greenpeace/WWF, 2005).

4. Jane Holden, *By Hook or by Crook: A Reference Manual on the Illegal Wildlife Trade and Prosecutions in the United Kingdom* (United Kingdom: The Royal Society for the Protection of Birds/WWF-UK/TRAFFIC International, 1998).

5. *Arborvitae: The IUCN/WWF Forest Conservation Newsletter* 32 (December 2006): 8, http://www.illegal-logging.info/uploads/arborvitae_32.pdf (accessed November 1, 2010).

6. Ibid.

7. Duncan Brack, Kevin Gray, and Gavin Hayman, *Controlling the International Trade in Illegally Logged Timber and Wood Products*, study prepared for the UK Department for International Development (London: Royal Institute of International Affairs, 2002); *Illegal Logging, Governance and Trade*.

8. *Arborvitae*, 2006.

9. *Arborvitae*, 2006; Charlie Pye-Smyth, *Logging in the Wild East: China and the Forest Crisis in the Russian Far East* (TRAFFIC, 2006).

10. "Data on the Black Market in Illegal Logging," Havocscope, http://www.havocscope.com/trafficking/logging.htm (accessed January 22, 2008).

11. Jamie Thomson and Ramzy Kanaan, *Conflict Timber: Dimensions of the Problem in Asia and Africa*, vol. 1 (Burlington, VT: ARD, Inc., 2004).

12. Brack et al., *Controlling the International Trade.*

13. Peter S. Goodman and Peter Finn, "Corruption Stains the Timber Trade: Forests Destroyed in China's Race to Feed Global Wood-Processing Industry," *Washington Post Foreign Service*, April 1, 2007, http://www.washingtonpost.com/wp-dyn/content/article/2007/03/31/AR2007033101287.html

14. Brack et al., *Controlling the International Trade.*

15. *Illegal Logging, Governance, and Trade.*

16. Ibid.

17. Simon Montlake, "Indonesia Battles Illegal Timber Trade," Special to the *Christian Science Monitor*, February 27, 2002, http://www.csmonitor.com/2002/0227/p07s01-woap.html (accessed October 4, 2010).

18. Ibid.

19. *Asia Pulp & Paper (APP) Threatens Bukit Tigapuluh Landscape: Report of Investigation Findings* (WWF Indonesia, 2008).

20. M. Richards, A. Wells, F. Del Gatto, A. Contreras-Hermosilla, and D. Pommier, "Impacts of Illegality and Barriers to Legality: A Diagnostic Analysis of Illegal Logging in Honduras and Nicaragua," *International Forestry Review* 5, no.3 (2003): 282–92.

21. Chris Fagan and Diego Shoobridge, *The Race for Peru's Last Mahogany Trees: Illegal Logging and the Alto Purus National Park*, (Round River Conservation Studies, 2007).

22. "Violent Mob Objects to Crackdown on Illegal Logging," *Tribune Review*, November 25, 2008, A2.

23. Richards et al., "Impacts of Illegality."

24. Ibid.

25. "International Crime Threat Assessment," *Trends in Organized Crime* 5, no. 4 (2000): 56–59.

26. *Exporting Destruction: Export Credits, Illegal Logging and Deforestation* (FERN, 2008), http://www.fern.org/media/documents/document_4155_4160.pdf (accessed October 4, 2010).

27. *Broken Promises: How World Bank Group Policies Fail to Protect Forests and Forest Peoples' Rights* (The Rainforest Foundation, 2005), http://www.forestpeoples.org/sites/fpp/files/publication/2010/08/wbbrokenpromisesapr05eng.pdf (accessed November 1, 2010).

28. Ibid.

29. Ibid.

30. Glastra, *Cut and Run*; Brack et al., *Controlling the International Trade.*

31. "The Envrionmental Effects of Illicit Crop Cultivation," in *The United Nations International Drug Control Programme, World Drug Report* (New York: Oxford University Press, 1997).

32. Richards et al., "Impacts of Illegality."

33. Ibid.

34. Ibid.

35. Ibid.

36. *Bulldozing Progress: Human Rights Abuses and Corruption in Papua New Guinea's Large Scale Logging Industry* (Australian Conservation Foundation, 2006), 3, http://www.acfonline.org.au/uploads/res/res_acf-celcor_full.pdf (accessed October 4, 2010).

37. Ibid.

38. Fagan and Shoobridge, *The Race for Peru's Last Mahogany Trees.*

39. Ibid.

40. Thomson and Kanaan, *Conflict Timber.*

41. Ibid.

42. Ibid.

43. Ibid.

44. Ibid.

45. Ibid.

46. Rina Agustini, Iola Leal Riesco, and Ridzki Rinanto Sigit, "Finding Solutions to Illegal Logging: Civil Society and the FLEGT Support Project" (2005), http://www.fern.org/sites/fern.org/files/media/documents/document _1287_1288.pdf (accessed November 1, 2010).

47. Ibid, 4.

48. Andrea Johnson, "U.S. Lacey Act: Respecting the Laws of Trade Partners," 2009, http://old.thejakartapost.com/detailededitorial.asp?fileid =20090305.F04&irec=7.

49. Jean-Marie Macabrey, "Deforestation: E.U. Committee Backs Broad New Penalties on Illegal Timber Trade," E&E Publishing, February 18, 2009, http://www.eenews.net/public/climatewire/2009/02/18/3 (accessed October 4, 2010).

50. Hewitt, *Failing the Forests.*

51. "WRM Bulletin 98" (January 9, 2005; edited March 18, 2009), http://www.illegal-logging.info/item_single.php?item=document&item_id=237 &approach_id=15 (accessed October 4, 2010).

52. Ibid.

53. VPAs are bilateral agreements between consumer and producer states in which both parties work in tandem to address illegal logging. Under FLEGT, EU customs agents could exclude timber imports from partner countries if the wood does not carry agreed upon certifications. *Arborvitae*, 4.

54. *Illegal Logging, Governance, and Trade.*

55. Hewitt, *Failing the Forests.*

56. Ibid.

57. *Illegal Logging, Governance, and Trade.*

58. *Arborvitae.*

59. *Arborvitae*; Dennis P. Dykstra, George Kuru, Rodney Taylor, Ruth Nussbaum, William Magrath, and Jane Story, *Technologies for Wood Tracking: Verification and Monitoring the Chain of Custody and Legal Compliance in the Timber Industry*, (Environmental and Social Development East Asia and Pacific Region Discussion Paper, World Bank, 2002).

60. *Illegal Logging, Governance, and Trade.*

61. *Arborvitae*.

62. *Arborvitae*, 13.

63. *Arborvitae*, 2.

64. *Arborvitae*.

65. "WRM Bulletin 98."

66. Chen Hin Keong, *The Role of CITES in Combating Illegal Logging: Current and Potential* (Cambridge: TRAFFIC International, 2006).

67. Ibid.

68. William M. Rhodes, Elizabeth P. Allen, and Myfanwy Callahan, *Illegal Logging: A Market-Based Analysis of Trafficking in Illegal Timber* (Cambridge, MA: Ibt Associates Inc., 2006).

CHAPTER 6

Conclusion

In recent decades, the illicit traffic in garbage and hazardous wastes, wildlife trafficking, IUU fishing, and illegal logging have exploded into a transnational crime problem, generating billions of dollars every year for illegal entrepreneurs and corrupt public officials. The negative environmental, social, and economic consequences of these activities are prodigious. Growing awareness of the problems and international conventions such as CITES and Basel (despite their limitations) are hopeful developments, but clearly the challenge of identifying and engineering the right balance of human and environmental interests remains. Enforcement and monitoring of environmental regulations and reductions in the various illicit traffics continue to be confounded by public corruption, certain aspects of global finance, and public regulatory policies that directly or indirectly facilitate crimes against the natural world and impoverished human populations. If growing public awareness and concern lie at the heart of concerted action, it is also true that economic disruptions like the global recession in 2008–2009 and purely human interests often trump environmental concerns. The key to responsible stewardship of the planet may depend on better educating people (consumers) to the fact that human interests and environmental interests are not mutually exclusive. More difficult will be the acknowledgement and acceptance that, in some cases, sacrifices that involve compromises from the competing interests may be necessary.

Events in 2009 and 2010 exemplify two issues that lie at the root of progress (or lack of it). The first concerns the issue of regulation, and the extent to which regulations may not only fail to curb problematic behaviors, but in many cases exacerbate them. In the case of the *Deepwater Horizon* oil disaster in the Gulf of Mexico, ostensibly business relationships and the regulatory framework were crafted in such a manner, whether purposeful or not, as to thwart environmental protection. The drilling rig was manufactured in South Korea, operated by a Swiss company under contract to a British oil company (British Petroleum), with significant responsibility for safety and environmental inspections falling to the tiny impoverished Pacific nation of the Marshall Islands—which outsourced many of its responsibilities to private companies! Principal regulatory authority fell to the U.S. Interior Department's Mine and Mineral Service (MMS), which, the year before the spill, gained some notoriety for figuratively—and in some cases literally—jumping in bed with entities it was charged with overseeing. The MMS noted but did nothing about hundreds of recent regulatory violations by British Petroleum.[1] Clearly, even well-crafted regulations are meaningless if the regulators are inept and/or corrupt. The second issue of concern is the already-noted problem of balancing economic prosperity versus environmental protection (and in some cases, the rights and claims of indigenous peoples), and is exemplified by a 2009 conflict between the Peruvian government and indigenous tribes. In a plan to secure a free trade agreement with the United States, Peru's administration of Alan Garcia lifted restrictions on exploration and development on mineral and timber-rich lands in parts of the Amazon forest. Indigenous tribes blocked roads for two months in a protest, and the government declared a state of emergency and suspended some constitutional protections. In June 2009, the tribes clashed with police, and 22 tribesmen and 18 police were killed.[2]

That conservation and environmental protection have not yet gained the level of importance granted to purely human concerns seems plain. For example, although it is a positive move, the regulation of garbage dumping by Caribbean nations has been astonishingly tardy: in November 2009, countries finally abolished rules that had permitted the disposal of metal, glass, and other refuse in the oceans a short distance from the shoreline, and practically any other trash farther out. Another example is the 2010 CITES convention in Doha, Qatar, where delegates failed to protect several declining species, including sharks hunted for their fins used in Chinese banquet soup (sharks take many years to reach adulthood, and in some fisheries, shark populations have declined by 80%–99%).[3] A U.S.-backed proposal at Doha to ban the export of Atlantic bluefin tuna was also rejected—a victory for Japan and dozens of poorer nations who argued the ban would decimate their economies.[4] Elsewhere, the traffic and irresponsible disposal of the ever-growing volume of electronic waste continues unabated. In India, countless small-time backyard and storefront electronic recyclers oppose a national law that would restrict recycling to licensed plants and ban the import of computers for "charity" or "reuse" (almost certainly transparent covers for international waste dumping). One unregulated independent operator complained to journalists that her skin itches and her head feels heavy from the processes involved in removing copper from discarded and burned circuit boards, but tolerated the employment of her 12-year old daughter in the industry, saying "I do not know if this is safe or unsafe, but no work is dirty if it feeds my family."[5] Again, competing interests exemplify the difficulty in implementing balanced and responsible approaches.

Wildlife poaching and trafficking also continues unabated.[6] The global ban on the ivory trade in 1989 has precipitated a lucrative black market, as the price of ivory has risen from $45 to $800 a pound over the last eight years (so even one tusk can be worth around $20,000). Sierra Leone lost its last elephants in December 2009, while fewer than 10 animals remain in Senegal. In Kenya, elephant-poaching deaths increased sevenfold from 2007 to 2010. A surge in demand for ivory in Asia could lead to additional declines in the population of African elephants.[7] Elsewhere, a global moratorium on commercial whaling since 1986 has failed to adequately protect many species. Japan, Norway, and Iceland routinely circumvent the commercial ban by abusing an allowance by the International Whaling Commission (IWC)—an 88-member body that lacks enforcement teeth—to conduct hunts for "scientific purposes." A plan to replace the moratorium with a strictly limited annual cull was scuttled in 2010 when the meeting of IWC members failed to reach agreement on hunts in Antarctic waters.[8]

While the problems are huge and seem intractable, some palliative measures have emerged. Perhaps the best way to reduce the traffic in hazardous waste and the illicit trade in fish, timber, and wildlife is to change patterns of consumer behavior, especially overconsumption in wealthier nations. Careful applications of taxes and subsidies can encourage desirable behaviors and discourage unattractive ones. For example, governmental units might follow the example set by the 2006 Plastic Debris Project Action Plan, where the state of California extended the redemption value of bottles and cans to plastics commonly found in the ocean. The initiative also provided low-interest loans to the fishing industry to help them develop more environmentally responsible practices.[9] Consumers can also be discouraged from using plastic bags at grocery stores. A heavy tax on plastic bags coupled with public support for companies that manufacture reusable cotton-string bags could eliminate a huge amount of plastic waste. In locations where this has been tried, relatively high costs to consumers who use plastic bags and enhanced profits for so-called "eco-bag" manufacturers discouraged the use of plastic bags; meanwhile, plastic grocery bag makers have an economic incentive to switch to the production of environmentally friendly bags.[10] In June 2010, the state assembly of California followed the lead of China and Bangladesh— where the use of plastic bags are prohibited—and approved a statewide ban that would require consumers who do not bring their own non-plastic bags to purchase from the grocery store reusable totes. Still, the countervailing interests remain: the head of the American Chemistry Council came out against the California ban, observing that consumers recovering from an economic recession would not welcome what amounts to a $1 billion tax on top of their grocery bills.[11]

While rapid technological development has contributed to environmental problems (for example, e-waste), technology can also present solutions. In some cases, technological advances can reduce the profitability and attractiveness of certain criminal enterprises. A good example is the traffic in garbage—if methods are developed that not only limit the volume of garbage, but also reduce the costs of responsible disposal, then *opportunities* for those who profit from the illegal dumping and traffic in waste would be inhibited. Some initiatives to reduce waste are surprisingly simple—the Greenshift Corporation developed an appliance that makes it easier for consumers to grasp the plastic lids used in fast food restaurants, thus reducing the *50 million pounds* of plastic lids that are discarded (unused) annually. In addition to reducing plastic waste, the appliance increases the profitability of each restaurant perhaps $20,000 a year in waste disposal costs, and conserves fossil-fuel resources used in the production of plastics.[12]

Igniting the profit motive of private companies and the desire of budget-conscious public officials to cut costs through technological advances may also inhibit the profitability of those illicit entrepreneurs who traffic in garbage and hazardous wastes. Formerly a costly burden to industry that bred illicit disposal, the recycling of PVC pipes has been transformed into a profitable commercial operation. In a process involving the thermal stabilization of PVC, companies like VEKA in Germany recycle entire windows containing PVC *without dismantling them*—a considerable cost-saving development. Solvay in Belgium recycles PVC bottles and pipes in such a way that the recycled materials retain properties equal to those of the original polymers.[13] Another example where technology provides an economic incentive involves a chemistry professor in New York who found a way to manufacture plastic that is easily converted to diesel fuel. In the process, DNA from a parasite is combined with E. coli bacteria to mass-produce the enzyme cutinase. Plastic waste is shredded, immersed in water, and combined with cutinase—three to five days later, biodiesel fuel floats to the top of the mixture. Since an individual soldier produces approximately seven pounds of packaging wastes per day, the Pentagon's Defense Advance Research Agency (DARPA) funded research into the cutinase production method, observing that processing the plastic waste could produce enough diesel to power an entire military base and save money in the event of relatively high crude oil prices.[14] In both the case of PVC recycling and the cutinase-based fuel production method, either profits are enhanced or costs are lessened while the volume of plastic waste is reduced—thereby limiting *opportunities* for illicit waste traffickers.

One clear impediment to progress is the intransigence of ideological positions. The Peruvian conflict described above exemplifies the need for a balanced approach, devoid of political ideologies and extreme viewpoints. Rapacious corporations and public officials who act in their own interests, not the public they serve, are no less helpful than unrealistic environmentalists who cry, "back to the Pleistoscene!" Nor should profit-making be eschewed as an absolute evil, as the wealth produced by private enterprise is the best antidote for poverty amelioration, not to mention the means by which environmental regulations and conservation measures are adequately funded.

Of course, even if proper regulatory balances are identified and codified for each of the various markets of concern, monitoring and enforcement present tremendous challenges. Clearly, an international effort will be required in which particular nations and regions look beyond their self-interests and adopt a global perspective—a politically unattractive and not-so-palatable approach for many governments and corporate entities. A system of powerful incentives and disincentives is obligatory if behaviors will change. Relatively wealthy consumer nations should continue to aid undeveloped nations in their efforts to police and enforce environmental laws, while the strings attached to foreign aid packages may need adjustment. Entities such as the World Bank and other financial institutions must consider a more balanced approach when it comes to financing some fishing and logging projects (loans for palm oil plantations in Indonesia being one prime example where investment capital directly contributes to environmental degradation, deforestation, and illegal logging).

Quite simply, the best way to lessen opportunities for criminals who traffic in hazardous wastes, illegally harvest fish, cut and traffic in timber in violation of the law, and perpetuate the illegal wildlife trade is for people to consume less of these services and products. Reduced demand translates directly into smaller black markets. Verification and certification of the various products and services will likewise enable consumers to make responsible choices.

Finally, environmental protection, human prosperity, and the suppression of the criminal enterprises discussed in this book come at a price—a balanced approach means that virtually everyone (including environmentalists) would necessarily make sacrifices, some products and services would likely become more expensive, and consumers would need to abstain from some products entirely. Intransigent ideological positions oriented around the poles of economic prosperity and unqualified conservation should be eschewed in deference to compromise. The adoption of elevated ethical standards for both private and public elites would foster an amelioration of the damages observed, and is in fact an indispensable key to success. At the most fundamental level, it must be recognized that the suppression of crimes against nature and a responsible stewardship of the planet will involve changing how we do business, how we live our lives, and how we think about our relationship with the natural world.

NOTES

1. "Massacre in Peru: A Dispatch on the Bloody Conflict," Banderas News, June 9, 2009, http://www.banderasnews.com/0906/edat-bloodyconflict.htm (accessed October 4, 2010).

2. "Tangled Global Web Cited in Spill," *Tribune Review*, June 15, 2010, A1.

3. "Wildlife Group Nixes Shark Protections," *Tribune Review*, March 24, 2010, A6.

4. "Bluefin Tuna, Polar Bear Denied Export Protections," *Tribune Review*, March 19, 2010, A2.

5. "India's E-Trash Safety Push Raises Concern," *Tribune Review*, June 13, 2010, A11.

6. "Data on the Black Market in Illegal Logging," Havocscope, http://www.havocscope.com/trafficking/logging.htm (accessed October 4, 2010); "Data on the Black Market in Wildlife and Animal Smuggling," http://www.havocscope.com/trafficking/wildlife.htm (accessed October 4, 2010); IllegalFishing.info, http://www.illegal-fishing.info (accessed October 4, 2010); "WFC Bulletin 98," Illegal-Fishing.info, http://www.illegal-logging.info/item_single.php?item=document&item_id=237&approach_id=15 (accessed October 4, 2010); http://www.traffic.org; http://www.worldwildlife.org.

7. "Asian Demand Has Elephants on Precipice," *Tribune Review*, June 16, 2010, A15.

8. Tom Pfeiffer, "Whaling Moratorium Talks Break Down," June 23, 2010, http://uk.reuters.com/article/idUKTRE65M26P20100623 (accessed November 1, 2010).

9. "Ocean Protection Council to Fight Marine Debris, Fund Low-interest Loans to Fishing Businesses and Communities," redOrbit, February 7, 2007, http://www.redorbit.com/news/science/831579/ocean_protection_council _to_fight_marine_debris_fund_lowinterest_loans/index.html (accessed October 4, 2010).

10. "Public Benefit Company Grows in Response to Concern About Plastic Bag Waste," redOrbit, March 14, 2006, http://www.redorbit.com/news/ science/427251/public_benefit_company_grows_in_response_to_concern _about_plastic/index.html (accessed October 4, 2010); "Enviro Brief: Degradable Plastic Waste Sacks," redOrbit, January 17, 2006, http://www .redorbit.com/news/science/359524/enviro_brief_degradable_plastic_waste _sacks/index.html (accessed October 4, 2010).

11. "California Advances Grocery Store Plastic Bag Ban," *Tribune* Review, June 4, 2010, A4.

12. "GreenShift Announces New Commercial Appliance for Reduction of Plastic Waste," redOrbit, July 18, 2005, http://www.redorbit.com/news/ technology/177818/greenshift_announces_new_commerical_appliance_for _reduction_of_plastic_waste/index.html (accessed October 4, 2010).

13. Matthew L. Wald, "Pentagon Puts Millions into Fuel for Plastics," redOrbit, April 9, 2007, http://www.redorbit.com/news/science/896830/ pentagon_puts_millions_into_fuel_from_plastic/index.html (accessed October 4, 2010).

14. Ibid.

Appendices

Effective international regulations will be integral in the effort to combat the crimes described in this book. Although their efficacy is debatable, three of the most influential laws intended to curtail the traffics in hazardous wastes and wildlife are the Basel Convention, CITES, and the Lacey Act.

The Basel Convention on the Control of Trans-boundary Movements of Hazardous Wastes and Their Disposal is a global environmental treaty that entered into force in 1992. There are 170 member countries (Parties). Under Basel, nations must be notified in advance of waste shipments, and the importers must consent to the shipments. Parties are required to enact domestic legislation to prevent and punish the illegal traffic in hazardous wastes. Unfortunately, flaws in the language of Basel have led to its easy circumvention. Radioactive waste is not covered, and key terms laying out rules and obligations of Parties are vaguely defined. Other limitations include weak prior informed consent rules, insufficient mechanisms for verification, and the fact that the Basel Secretariat has no power to monitor Parties or apply sanctions. The text of the Basel Convention is provided in Appendix A.

The Convention on the International Trade in Endangered Species of Wild Fauna and Flora (CITES) is the most significant international treaty used to combat wildlife trafficking. Under CITES, protection is afforded to 33,000 species of animals and plants. There are 172 Member Parties. Import, export, reexport, and introduction from the sea of listed species are subject to controls through licensing. Parties must designate one or more Management Authorities in charge of administering the licensing system and one or more Scientific Authorities to give advice on the effects of the trade on species. Import and export of species covered by CITES is permissible only when proper documentation is obtained and displayed at ports of entry or exit. Species are divided into three categories, with different levels of protections and rules pertaining to each (95% of CITES-listed species are not endangered). While ambitious, CITES is limited—it does not have enforcement powers, nor does it issue permits. Management authorities in Party countries are responsible for regulation, and many simply lack the resources and/or political will to effectively enforce regulations. The full text of CITES is provided in Appendix B.

The Lacey Act, along with the Endangered Species Act, is the principal tool by which authorities enforce laws meant to protect wildlife in the United States. The law covers both interstate and foreign commerce, and provides penalties for traffickers that could include five year prison sentences. The text of the Lacey Act is in Appendix C.

Appendix A

Basel Convention on the Control of Transboundary Movements of Hazardous Wastes and Their Disposal[1]

PREAMBLE

The Parties to this Convention,

Aware of the risk of damage to human health and the environment caused by hazardous wastes and other wastes and the transboundary movement thereof,

Mindful of the growing threat to human health and the environment posed by the increased generation and complexity, and transboundary movement of hazardous wastes and other wastes,

Mindful also that the most effective way of protecting human health and the environment from the dangers posed by such wastes is the reduction of their generation to a minimum in terms of quantity and/or hazard potential,

Convinced that States should take necessary measures to ensure that the management of hazardous wastes and other wastes including their transboundary movement and disposal is consistent with the protection of human health and the environment whatever the place of disposal,

Noting that States should ensure that the generator should carry out duties with regard to the transport and disposal of hazardous wastes and other wastes in a manner that is consistent with the protection of the environment, whatever the place of disposal,

---

[1]The present text incorporates amendments to the Convention adopted subsequent to its entry into force and that are in force as at 8 October 2005. Only the text of the Convention as kept in the custody of the Secretary-General of the United Nations in his capacity as Depositary constitutes the authentic version of the Convention, as modified by any amendments and/or corrections thereto. This publication is issued for information purposes only.

Fully recognizing that any State has the sovereign right to ban the entry or disposal of foreign hazardous wastes and other wastes in its territory,

Recognizing also the increasing desire for the prohibition of transboundary movements of hazardous wastes and their disposal in other States, especially developing countries,

Convinced that hazardous wastes and other wastes should, as far as is compatible with environmentally sound and efficient management, be disposed of in the State where they were generated,

Aware also that transboundary movements of such wastes from the State of their generation to any other State should be permitted only when conducted under conditions which do not endanger human health and the environment, and under conditions in conformity with the provisions of this Convention,

Considering that enhanced control of transboundary movement of hazardous wastes and other wastes will act as an incentive for their environmentally sound management and for the reduction of the volume of such transboundary movement,

Convinced that States should take measures for the proper exchange of information on and control of the transboundary movement of hazardous wastes and other wastes from and to those States,

Noting that a number of international and regional agreements have addressed the issue of protection and preservation of the environment with regard to the transit of dangerous goods,

Taking into account the Declaration of the United Nations Conference on the Human Environment (Stockholm, 1972), the Cairo Guidelines and Principles for the Environmentally Sound Management of Hazardous Wastes adopted by the Governing Council of the United Nations Environment Programme (UNEP) by decision 14/30 of 17 June 1987, the Recommendations of the United Nations Committee of Experts on the Transport of Dangerous Goods (formulated in 1957 and updated biennially), relevant recommendations, declarations, instruments and regulations adopted within the United Nations system and the work and studies done within other international and regional organizations,

Mindful of the spirit, principles, aims and functions of the World Charter for Nature adopted by the General Assembly of the United Nations at its thirty-seventh session (1982) as the rule of ethics in respect of the protection of the human environment and the conservation of natural resources,

Affirming that States are responsible for the fulfilment of their international obligations concerning the protection of human health and protection and preservation of the environment, and are liable in accordance with international law,

Recognizing that in the case of a material breach of the provisions of this Convention or any protocol thereto the relevant international law of treaties shall apply,

Aware of the need to continue the development and implementation of environmentally sound low-waste technologies, recycling options, good house-keeping and management systems with a view to reducing to a minimum the generation of hazardous wastes and other wastes,

Aware also of the growing international concern about the need for stringent control of transboundary movement of hazardous wastes and other wastes, and of the need as far as possible to reduce such movement to a minimum,

Concerned about the problem of illegal transboundary traffic in hazardous wastes and other wastes,

Taking into account also the limited capabilities of the developing countries to manage hazardous wastes and other wastes,

Recognizing the need to promote the transfer of technology for the sound management of hazardous wastes and other wastes produced locally, particularly to the developing countries in accordance with the spirit of the Cairo Guidelines and decision 14/16 of the Governing Council of UNEP on Promotion of the transfer of environmental protection technology,

Recognizing also that hazardous wastes and other wastes should be transported in accordance with relevant international conventions and recommendations,

Convinced also that the transboundary movement of hazardous wastes and other wastes should be permitted only when the transport and the ultimate disposal of such wastes is environmentally sound, and

Determined to protect, by strict control, human health and the environment against the adverse effects which may result from the generation and management of hazardous wastes and other wastes,

HAVE AGREED AS FOLLOWS:

ARTICLE 1

SCOPE OF THE CONVENTION

1. The following wastes that are subject to transboundary movement shall be "hazardous wastes" for the purposes of this Convention:

    (a) Wastes that belong to any category contained in Annex I, unless they do not possess any of the characteristics contained in Annex III; and

(b) Wastes that are not covered under paragraph (a) but are defined as, or are considered to be, hazardous wastes by the domestic legislation of the Party of export, import or transit.

2. Wastes that belong to any category contained in Annex II that are subject to transboundary movement shall be "other wastes" for the purposes of this Convention.

3. Wastes which, as a result of being radioactive, are subject to other international control systems, including international instruments, applying specifically to radioactive materials, are excluded from the scope of this Convention.

4. Wastes which derive from the normal operations of a ship, the discharge of which is covered by another international instrument, are excluded from the scope of this Convention.

ARTICLE 2

DEFINITIONS

For the purposes of this Convention:

1. "Wastes" are substances or objects which are disposed of or are intended to be disposed of or are required to be disposed of by the provisions of national law;

2. "Management" means the collection, transport and disposal of hazardous wastes or other wastes, including after-care of disposal sites;

3. "Transboundary movement" means any movement of hazardous wastes or other wastes from an area under the national jurisdiction of one State to or through an area under the national jurisdiction of another State or to or through an area not under the national jurisdiction of any State, provided at least two States are involved in the movement;

4. "Disposal" means any operation specified in Annex IV to this Convention;

5. "Approved site or facility" means a site or facility for the disposal of hazardous wastes or other wastes which is authorized or permitted to operate for this purpose by a relevant authority of the State where the site or facility is located;

6. "Competent authority" means one governmental authority designated by a Party to be responsible, within such geographical areas as the Party may think fit, for receiving the notification of a transboundary movement of hazardous wastes or other wastes, and any information related to it, and for responding to such a notification, as provided in Article 6;

7. "Focal point" means the entity of a Party referred to in Article 5 responsible for receiving and submitting information as provided for in Articles 13 and 16;

8. "Environmentally sound management of hazardous wastes or other wastes" means taking all practicable steps to ensure that hazardous wastes or other wastes are managed in a manner which will protect human health and the environment against the adverse effects which may result from such wastes;

9. "Area under the national jurisdiction of a State" means any land, marine area or airspace within which a State exercises administrative and regulatory responsibility in

accordance with international law in regard to the protection of human health or the environment;

10. "State of export" means a Party from which a transboundary movement of hazardous wastes or other wastes is planned to be initiated or is initiated;

11. "State of import" means a Party to which a transboundary movement of hazardous wastes or other wastes is planned or takes place for the purpose of disposal therein or for the purpose of loading prior to disposal in an area not under the national jurisdiction of any State;

12. "State of transit" means any State, other than the State of export or import, through which a movement of hazardous wastes or other wastes is planned or takes place;

13. "States concerned" means Parties which are States of export or import, or transit States, whether or not Parties;

14. "Person" means any natural or legal person;

15. "Exporter" means any person under the jurisdiction of the State of export who arranges for hazardous wastes or other wastes to be exported;

16. "Importer" means any person under the jurisdiction of the State of import who arranges for hazardous wastes or other wastes to be imported;

17. "Carrier" means any person who carries out the transport of hazardous wastes or other wastes;

18. "Generator" means any person whose activity produces hazardous wastes or other wastes or, if that person is not known, the person who is in possession and/or control of those wastes;

19. "Disposer" means any person to whom hazardous wastes or other wastes are shipped and who carries out the disposal of such wastes;

20. "Political and/or economic integration organization" means an organization constituted by sovereign States to which its member States have transferred competence in respect of matters governed by this Convention and which has been duly authorized, in accordance with its internal procedures, to sign, ratify, accept, approve, formally confirm or accede to it;

21. "Illegal traffic" means any transboundary movement of hazardous wastes or other wastes as specified in Article 9.

ARTICLE 3
NATIONAL DEFINITIONS OF HAZARDOUS WASTES

1. Each Party shall, within six months of becoming a Party to this Convention, inform the Secretariat of the Convention of the wastes, other than those listed in Annexes I and II, considered or defined as hazardous under its national legislation and of any requirements concerning transboundary movement procedures applicable to such wastes.

2. Each Party shall subsequently inform the Secretariat of any significant changes to the information it has provided pursuant to paragraph 1.

3. The Secretariat shall forthwith inform all Parties of the information it has received pursuant to paragraphs 1 and 2.

4. Parties shall be responsible for making the information transmitted to them by the Secretariat under paragraph 3 available to their exporters.

ARTICLE 4[2]

GENERAL OBLIGATIONS

1. (a) Parties exercising their right to prohibit the import of hazardous wastes or other wastes for disposal shall inform the other Parties of their decision pursuant to Article 13.

   (b) Parties shall prohibit or shall not permit the export of hazardous wastes and other wastes to the Parties which have prohibited the import of such wastes, when notified pursuant to subparagraph (a) above.

   (c) Parties shall prohibit or shall not permit the export of hazardous wastes and other wastes if the State of import does not consent in writing to the specific import, in the case where that State of import has not prohibited the import of such wastes.

2. Each Party shall take the appropriate measures to:

   (a) Ensure that the generation of hazardous wastes and other wastes within it is reduced to a minimum, taking into account social, technological and economic aspects;

   (b) Ensure the availability of adequate disposal facilities, for the environmentally sound management of hazardous wastes and other wastes, that shall be located, to the extent possible, within it, whatever the place of their disposal;

---

[2]The Conference of the Parties adopted Decision III/1 at its third meeting to amend the Convention by adding, *inter alia*, a new Article 4A. The amendment is not yet in force. The relevant part of Decision III/1 provides as follows:

"The Conference

. . .

3. Decides to adopt the following amendment to the Convention:

. . .

"Insert new Article 4A:

1. Each Party listed in Annex VII shall prohibit all transboundary movements of hazardous wastes which are destined for operations according to Annex IV A, to States not listed in Annex VII.

2. Each Party listed in Annex VII shall phase out by 31 December 1997, and prohibit as of that date, all transboundary movements of hazardous wastes under Article 1(1)(a) of the Convention which are destined for operations according to Annex IV B to States not listed in Annex VII. Such transboundary movement shall not be prohibited unless the wastes in question are characterised as hazardous under the Convention. . . . ' "

(c) Ensure that persons involved in the management of hazardous wastes or other wastes within it take such steps as are necessary to prevent pollution due to hazardous wastes and other wastes arising from such management and, if such pollution occurs, to minimize the consequences thereof for human health and the environment;

(d) Ensure that the transboundary movement of hazardous wastes and other wastes is reduced to the minimum consistent with the environmentally sound and efficient management of such wastes, and is conducted in a manner which will protect human health and the environment against the adverse effects which may result from such movement;

(e) Not allow the export of hazardous wastes or other wastes to a State or group of States belonging to an economic and/or political integration organization that are Parties, particularly developing countries, which have prohibited by their legislation all imports, or if it has reason to believe that the wastes in question will not be managed in an environmentally sound manner, according to criteria to be decided on by the Parties at their first meeting;

(f) Require that information about a proposed transboundary movement of hazardous wastes and other wastes be provided to the States concerned, according to Annex V A, to state clearly the effects of the proposed movement on human health and the environment;

(g) Prevent the import of hazardous wastes and other wastes if it has reason to believe that the wastes in question will not be managed in an environmentally sound manner;

(h) Co-operate in activities with other Parties and interested organizations, directly and through the Secretariat, including the dissemination of information on the transboundary movement of hazardous wastes and other wastes, in order to improve the environmentally sound management of such wastes and to achieve the prevention of illegal traffic.

3. The Parties consider that illegal traffic in hazardous wastes or other wastes is criminal.

4. Each Party shall take appropriate legal, administrative and other measures to implement and enforce the provisions of this Convention, including measures to prevent and punish conduct in contravention of the Convention.

5. A Party shall not permit hazardous wastes or other wastes to be exported to a non-Party or to be imported from a non-Party.

6. The Parties agree not to allow the export of hazardous wastes or other wastes for disposal within the area south of 60° South latitude, whether or not such wastes are subject to transboundary movement.

7. Furthermore, each Party shall:

(a) Prohibit all persons under its national jurisdiction from transporting or disposing of hazardous wastes or other wastes unless such persons are authorized or allowed to perform such types of operations;

(b) Require that hazardous wastes and other wastes that are to be the subject of a transboundary movement be packaged, labelled, and transported in conformity with generally accepted and recognized international rules and standards in the field of packaging, labelling, and transport, and that due account is taken of relevant internationally recognized practices;

(c) Require that hazardous wastes and other wastes be accompanied by a movement document from the point at which a transboundary movement commences to the point of disposal.

8. Each Party shall require that hazardous wastes or other wastes, to be exported, are managed in an environmentally sound manner in the State of import or elsewhere. Technical guidelines for the environmentally sound management of wastes subject to this Convention shall be decided by the Parties at their first meeting.

9. Parties shall take the appropriate measures to ensure that the transboundary movement of hazardous wastes and other wastes only be allowed if:

(a) The State of export does not have the technical capacity and the necessary facilities, capacity or suitable disposal sites in order to dispose of the wastes in question in an environmentally sound and efficient manner; or

(b) The wastes in question are required as a raw material for recycling or recovery industries in the State of import; or

(c) The transboundary movement in question is in accordance with other criteria to be decided by the Parties, provided those criteria do not differ from the objectives of this Convention.

10. The obligation under this Convention of States in which hazardous wastes and other wastes are generated to require that those wastes are managed in an environmentally sound manner may not under any circumstances be transferred to the States of import or transit.

11. Nothing in this Convention shall prevent a Party from imposing additional requirements that are consistent with the provisions of this Convention, and are in accordance with the rules of international law, in order better to protect human health and the environment.

12. Nothing in this Convention shall affect in any way the sovereignty of States over their territorial sea established in accordance with international law, and the sovereign rights and the jurisdiction which States have in their exclusive economic zones and their continental shelves in accordance with international law, and the exercise by ships and aircraft of all States of navigational rights and freedoms as provided for in international law and as reflected in relevant international instruments.

13. Parties shall undertake to review periodically the possibilities for the reduction of the amount and/or the pollution potential of hazardous wastes and other wastes which are exported to other States, in particular to developing countries.

ARTICLE 5

DESIGNATION OF COMPETENT AUTHORITIES AND FOCAL POINT

To facilitate the implementation of this Convention, the Parties shall:

1. Designate or establish one or more competent authorities and one focal point. One competent authority shall be designated to receive the notification in case of a State of transit.

2. Inform the Secretariat, within three months of the date of the entry into force of this Convention for them, which agencies they have designated as their focal point and their competent authorities.

3. Inform the Secretariat, within one month of the date of decision, of any changes regarding the designation made by them under paragraph 2 above.

ARTICLE 6
Transboundary Movement between Parties

1. The State of export shall notify, or shall require the generator or exporter to notify, in writing, through the channel of the competent authority of the State of export, the competent authority of the States concerned of any proposed transboundary movement of hazardous wastes or other wastes. Such notification shall contain the declarations and information specified in Annex V A, written in a language acceptable to the State of import. Only one notification needs to be sent to each State concerned.

2. The State of import shall respond to the notifier in writing, consenting to the movement with or without conditions, denying permission for the movement, or requesting additional information. A copy of the final response of the State of import shall be sent to the competent authorities of the States concerned which are Parties.

3. The State of export shall not allow the generator or exporter to commence the transboundary movement until it has received written confirmation that:

   (a) The notifier has received the written consent of the State of import; and

   (b) The notifier has received from the State of import confirmation of the existence of a contract between the exporter and the disposer specifying environmentally sound management of the wastes in question.

4. Each State of transit which is a Party shall promptly acknowledge to the notifier receipt of the notification. It may subsequently respond to the notifier in writing, within 60 days, consenting to the movement with or without conditions, denying permission for the movement, or requesting additional information. The State of export shall not allow the transboundary movement to commence until it has received the written consent of the State of transit. However, if at any time a Party decides not to require prior written consent, either generally or under specific conditions, for transit transboundary movements of hazardous wastes or other wastes, or modifies its requirements in this respect, it shall forthwith inform the other Parties of its decision pursuant to Article 13. In this latter case, if no response is received by the State of export within 60 days of the receipt of a given notification by the State of transit, the State of export may allow the export to proceed through the State of transit.

5. In the case of a transboundary movement of wastes where the wastes are legally defined as or considered to be hazardous wastes only:

   (a) By the State of export, the requirements of paragraph 9 of this Article that apply to the importer or disposer and the State of import shall apply *mutatis mutandis* to the exporter and State of export, respectively;

(b) By the State of import, or by the States of import and transit which are Parties, the requirements of paragraphs 1, 3, 4 and 6 of this Article that apply to the exporter and State of export shall apply *mutatis mutandis* to the importer or disposer and State of import, respectively; or

(c) By any State of transit which is a Party, the provisions of paragraph 4 shall apply to such State.

6. The State of export may, subject to the written consent of the States concerned, allow the generator or the exporter to use a general notification where hazardous wastes or other wastes having the same physical and chemical characteristics are shipped regularly to the same disposer via the same customs office of exit of the State of export via the same customs office of entry of the State of import, and, in the case of transit, via the same customs office of entry and exit of the State or States of transit.

7. The States concerned may make their written consent to the use of the general notification referred to in paragraph 6 subject to the supply of certain information, such as the exact quantities or periodical lists of hazardous wastes or other wastes to be shipped.

8. The general notification and written consent referred to in paragraphs 6 and 7 may cover multiple shipments of hazardous wastes or other wastes during a maximum period of 12 months.

9. The Parties shall require that each person who takes charge of a transboundary movement of hazardous wastes or other wastes sign the movement document either upon delivery or receipt of the wastes in question. They shall also require that the disposer inform both the exporter and the competent authority of the State of export of receipt by the disposer of the wastes in question and, in due course, of the completion of disposal as specified in the notification. If no such information is received within the State of export, the competent authority of the State of export or the exporter shall so notify the State of import.

10. The notification and response required by this Article shall be transmitted to the competent authority of the Parties concerned or to such governmental authority as may be appropriate in the case of non-Parties.

11. Any transboundary movement of hazardous wastes or other wastes shall be covered by insurance, bond or other guarantee as may be required by the State of import or any State of transit which is a Party.

ARTICLE 7

TRANSBOUNDARY MOVEMENT FROM A PARTY THROUGH STATES WHICH ARE NOT PARTIES

Paragraph 1 of Article 6 of the Convention shall apply *mutatis mutandis* to transboundary movement of hazardous wastes or other wastes from a Party through a State or States which are not Parties.

ARTICLE 8

Duty to Re-import

When a transboundary movement of hazardous wastes or other wastes to which the consent of the States concerned has been given, subject to the provisions of this Convention, cannot be completed in accordance with the terms of the contract, the State of export shall ensure that the wastes in question are taken back into the State of export, by the exporter, if alternative arrangements cannot be made for their disposal in an environmentally sound manner, within 90 days from the time that the importing State informed the State of export and the Secretariat, or such other period of time as the States concerned agree. To this end, the State of export and any Party of transit shall not oppose, hinder or prevent the return of those wastes to the State of export.

ARTICLE 9

Illegal Traffic

1. For the purpose of this Convention, any transboundary movement of hazardous wastes or other wastes:
   (a) without notification pursuant to the provisions of this Convention to all States concerned; or
   (b) without the consent pursuant to the provisions of this Convention of a State concerned; or
   (c) with consent obtained from States concerned through falsification, misrepresentation or fraud; or
   (d) that does not conform in a material way with the documents; or
   (e) that results in deliberate disposal (e.g., dumping) of hazardous wastes or other wastes in contravention of this Convention and of general principles of international law,
   shall be deemed to be illegal traffic.

2. In case of a transboundary movement of hazardous wastes or other wastes deemed to be illegal traffic as the result of conduct on the part of the exporter or generator, the State of export shall ensure that the wastes in question are:
   (a) taken back by the exporter or the generator or, if necessary, by itself into the State of export, or, if impracticable,
   (b) are otherwise disposed of in accordance with the provisions of this Convention,
   within 30 days from the time the State of export has been informed about the illegal traffic or such other period of time as States concerned may agree.

To this end the Parties concerned shall not oppose, hinder or prevent the return of those wastes to the State of export.

3. In the case of a transboundary movement of hazardous wastes or other wastes deemed to be illegal traffic as the result of conduct on the part of the importer or disposer, the State of import shall ensure that the wastes in question are disposed of in an environmentally sound manner by the importer or disposer or, if necessary, by itself within 30 days from the time the illegal traffic has come to the attention of the State of import or such other period of time as the States concerned may agree. To this end, the Parties concerned shall co-operate, as necessary, in the disposal of the wastes in an environmentally sound manner.

4. In cases where the responsibility for the illegal traffic cannot be assigned either to the exporter or generator or to the importer or disposer, the Parties concerned or other Parties, as appropriate, shall ensure, through co-operation, that the wastes in question are disposed of as soon as possible in an environmentally sound manner either in the State of export or the State of import or elsewhere as appropriate.

5. Each Party shall introduce appropriate national/domestic legislation to prevent and punish illegal traffic. The Parties shall co-operate with a view to achieving the objects of this Article.

ARTICLE 10

INTERNATIONAL CO-OPERATION

1. The Parties shall co-operate with each other in order to improve and achieve environmentally sound management of hazardous wastes and other wastes.

2. To this end, the Parties shall:

    (a) Upon request, make available information, whether on a bilateral or multilateral basis, with a view to promoting the environmentally sound management of hazardous wastes and other wastes, including harmonization of technical standards and practices for the adequate management of hazardous wastes and other wastes;

    (b) Co-operate in monitoring the effects of the management of hazardous wastes on human health and the environment;

    (c) Co-operate, subject to their national laws, regulations and policies, in the development and implementation of new environmentally sound low-waste technologies and the improvement of existing technologies with a view to eliminating, as far as practicable, the generation of hazardous wastes and other wastes and achieving more effective and efficient methods of ensuring their management in an environmentally sound manner, including the study of the economic, social and environmental effects of the adoption of such new or improved technologies;

    (d) Co-operate actively, subject to their national laws, regulations and policies, in the transfer of technology and management systems related to the environmentally sound management of hazardous wastes and other wastes.

They shall also co-operate in developing the technical capacity among Parties, especially those which may need and request technical assistance in this field;

(e) Co-operate in developing appropriate technical guidelines and/or codes of practice.

3. The Parties shall employ appropriate means to co-operate in order to assist developing countries in the implementation of subparagraphs a, b, c and d of paragraph 2 of Article 4.

4. Taking into account the needs of developing countries, co-operation between Parties and the competent international organizations is encouraged to promote, *inter alia*, public awareness, the development of sound management of hazardous wastes and other wastes and the adoption of new low-waste technologies.

ARTICLE 11

BILATERAL, MULTILATERAL AND REGIONAL AGREEMENTS

1. Notwithstanding the provisions of Article 4 paragraph 5, Parties may enter into bilateral, multilateral, or regional agreements or arrangements regarding transboundary movement of hazardous wastes or other wastes with Parties or non-Parties provided that such agreements or arrangements do not derogate from the environmentally sound management of hazardous wastes and other wastes as required by this Convention. These agreements or arrangements shall stipulate provisions which are not less environmentally sound than those provided for by this Convention in particular taking into account the interests of developing countries.

2. Parties shall notify the Secretariat of any bilateral, multilateral or regional agreements or arrangements referred to in paragraph 1 and those which they have entered into prior to the entry into force of this Convention for them, for the purpose of controlling transboundary movements of hazardous wastes and other wastes which take place entirely among the Parties to such agreements. The provisions of this Convention shall not affect transboundary movements which take place pursuant to such agreements provided that such agreements are compatible with the environmentally sound management of hazardous wastes and other wastes as required by this Convention.

ARTICLE 12

CONSULTATIONS ON LIABILITY

The Parties shall co-operate with a view to adopting, as soon as practicable, a protocol setting out appropriate rules and procedures in the field of liability and compensation for damage resulting from the transboundary movement and disposal of hazardous wastes and other wastes.

ARTICLE 13

TRANSMISSION OF INFORMATION

1. The Parties shall, whenever it comes to their knowledge, ensure that, in the case of an accident occurring during the transboundary movement of hazardous wastes or other wastes or their disposal, which are likely to present risks to human health and the environment in other States, those States are immediately informed.

2. The Parties shall inform each other, through the Secretariat, of:

   (a) Changes regarding the designation of competent authorities and/or focal points, pursuant to Article 5;

   (b) Changes in their national definition of hazardous wastes, pursuant to Article 3;

   and, as soon as possible,

   (c) Decisions made by them not to consent totally or partially to the import of hazardous wastes or other wastes for disposal within the area under their national jurisdiction;

   (d) Decisions taken by them to limit or ban the export of hazardous wastes or other wastes;

   (e) Any other information required pursuant to paragraph 4 of this Article.

3. The Parties, consistent with national laws and regulations, shall transmit, through the Secretariat, to the Conference of the Parties established under Article 15, before the end of each calendar year, a report on the previous calendar year, containing the following information:

   (a) Competent authorities and focal points that have been designated by them pursuant to Article 5;

   (b) Information regarding transboundary movements of hazardous wastes or other wastes in which they have been involved, including:

      (i) The amount of hazardous wastes and other wastes exported, their category, characteristics, destination, any transit country and disposal method as stated on the response to notification;

      (ii) The amount of hazardous wastes and other wastes imported, their category, characteristics, origin, and disposal methods;

      (iii) Disposals which did not proceed as intended;

      (iv) Efforts to achieve a reduction of the amount of hazardous wastes or other wastes subject to transboundary movement;

   (c) Information on the measures adopted by them in implementation of this Convention;

   (d) Information on available qualified statistics which have been compiled by them on the effects on human health and the environment of the generation, transportation and disposal of hazardous wastes or other wastes;

   (e) Information concerning bilateral, multilateral and regional agreements and arrangements entered into pursuant to Article 11 of this Convention;

(f) Information on accidents occurring during the transboundary movement and disposal of hazardous wastes and other wastes and on the measures undertaken to deal with them;

(g) Information on disposal options operated within the area of their national jurisdiction;

(h) Information on measures undertaken for development of technologies for the reduction and/or elimination of production of hazardous wastes and other wastes; and

(i) Such other matters as the Conference of the Parties shall deem relevant.

4. The Parties, consistent with national laws and regulations, shall ensure that copies of each notification concerning any given transboundary movement of hazardous wastes or other wastes, and the response to it, are sent to the Secretariat when a Party considers that its environment may be affected by that transboundary movement has requested that this should be done.

ARTICLE 14

FINANCIAL ASPECTS

1. The Parties agree that, according to the specific needs of different regions and subregions, regional or sub-regional centres for training and technology transfers regarding the management of hazardous wastes and other wastes and the minimization of their generation should be established. The Parties shall decide on the establishment of appropriate funding mechanisms of a voluntary nature.

2. The Parties shall consider the establishment of a revolving fund to assist on an interim basis in case of emergency situations to minimize damage from accidents arising from transboundary movements of hazardous wastes and other wastes or during the disposal of those wastes.

ARTICLE 15

CONFERENCE OF THE PARTIES

1. A Conference of the Parties is hereby established. The first meeting of the Conference of the Parties shall be convened by the Executive Director of UNEP not later than one year after the entry into force of this Convention. Thereafter, ordinary meetings of the Conference of the Parties shall be held at regular intervals to be determined by the Conference at its first meeting.

2. Extraordinary meetings of the Conference of the Parties shall be held at such other times as may be deemed necessary by the Conference, or at the written request of any Party, provided that, within six months of the request being communicated to them by the Secretariat, it is supported by at least one third of the Parties.

3. The Conference of the Parties shall by consensus agree upon and adopt rules of procedure for itself and for any subsidiary body it may establish, as well as

financial rules to determine in particular the financial participation of the Parties under this Convention.

4. The Parties at their first meeting shall consider any additional measures needed to assist them in fulfilling their responsibilities with respect to the protection and the preservation of the marine environment in the context of this Convention.

5. The Conference of the Parties shall keep under continuous review and evaluation the effective implementation of this Convention, and, in addition, shall:

   (a) Promote the harmonization of appropriate policies, strategies and measures for minimizing harm to human health and the environment by hazardous wastes and other wastes;

   (b) Consider and adopt, as required, amendments to this Convention and its annexes, taking into consideration, *inter alia*, available scientific, technical, economic and environmental information;

   (c) Consider and undertake any additional action that may be required for the achievement of the purposes of this Convention in the light of experience gained in its operation and in the operation of the agreements and arrangements envisaged in Article 11;

   (d) Consider and adopt protocols as required; and

   (e) Establish such subsidiary bodies as are deemed necessary for the implementation of this Convention.

6. The United Nations, its specialized agencies, as well as any State not Party to this Convention, may be represented as observers at meetings of the Conference of the Parties. Any other body or agency, whether national or international, governmental or non-governmental, qualified in fields relating to hazardous wastes or other wastes which has informed the Secretariat of its wish to be represented as an observer at a meeting of the Conference of the Parties, may be admitted unless at least one third of the Parties present object. The admission and participation of observers shall be subject to the rules of procedure adopted by the Conference of the Parties.

7. The Conference of the Parties shall undertake three years after the entry into force of this Convention, and at least every six years thereafter, an evaluation of its effectiveness and, if deemed necessary, to consider the adoption of a complete or partial ban of transboundary movements of hazardous wastes and other wastes in light of the latest scientific, environmental, technical and economic information.

ARTICLE 16

SECRETARIAT

1. The functions of the Secretariat shall be:

   (a) To arrange for and service meetings provided for in Articles 15 and 17;

   (b) To prepare and transmit reports based upon information received in accordance with Articles 3, 4, 6, 11 and 13 as well as upon information derived from meetings of subsidiary bodies established under Article 15 as well as upon, as appropriate, information provided by relevant intergovernmental and non-governmental entities;

(c) To prepare reports on its activities carried out in implementation of its functions under this Convention and present them to the Conference of the Parties;

(d) To ensure the necessary coordination with relevant international bodies, and in particular to enter into such administrative and contractual arrangements as may be required for the effective discharge of its function;

(e) To communicate with focal points and competent authorities established by the Parties in accordance with Article 5 of this Convention;

(f) To compile information concerning authorized national sites and facilities of Parties available for the disposal of their hazardous wastes and other wastes and to circulate this information among Parties;

(g) To receive and convey information from and to Parties on:

– sources of technical assistance and training;

– available technical and scientific know-how;

– sources of advice and expertise; and

– availability of resources

with a view to assisting them, upon request, in such areas as:

– the handling of the notification system of this Convention;

– the management of hazardous wastes and other wastes;

– environmentally sound technologies relating to hazardous wastes and other wastes; such as low- and non-waste technology;

– the assessment of disposal capabilities and sites;

– the monitoring of hazardous wastes and other wastes; and

– emergency responses;

(h) To provide Parties, upon request, with information on consultants or consulting firms having the necessary technical competence in the field, which can assist them to examine a notification for a transboundary movement, the concurrence of a shipment of hazardous wastes or other wastes with the relevant notification, and/or the fact that the proposed disposal facilities for hazardous wastes or other wastes are environmentally sound, when they have reason to believe that the wastes in question will not be managed in an environmentally sound manner. Any such examination would not be at the expense of the Secretariat;

(i) To assist Parties upon request in their identification of cases of illegal traffic and to circulate immediately to the Parties concerned any information it has received regarding illegal traffic;

(j) To co-operate with Parties and with relevant and competent international organizations and agencies in the provision of experts and equipment for the purpose of rapid assistance to States in the event of an emergency situation; and

(k) To perform such other functions relevant to the purposes of this Convention as may be determined by the Conference of the Parties.

2. The secretariat functions will be carried out on an interim basis by UNEP until the completion of the first meeting of the Conference of the Parties held pursuant to Article 15.

3. At its first meeting, the Conference of the Parties shall designate the Secretariat from among those existing competent intergovernmental organizations which have signified their willingness to carry out the secretariat functions under this Convention. At this meeting, the Conference of the Parties shall also evaluate the implementation by the interim Secretariat of the functions assigned to it, in particular under paragraph 1 above, and decide upon the structures appropriate for those functions.

ARTICLE 17

AMENDMENT OF THE CONVENTION

1. Any Party may propose amendments to this Convention and any Party to a protocol may propose amendments to that protocol. Such amendments shall take due account, *inter alia*, of relevant scientific and technical considerations.

2. Amendments to this Convention shall be adopted at a meeting of the Conference of the Parties. Amendments to any protocol shall be adopted at a meeting of the Parties to the protocol in question. The text of any proposed amendment to this Convention or to any protocol, except as may otherwise be provided in such protocol, shall be communicated to the Parties by the Secretariat at least six months before the meeting at which it is proposed for adoption. The Secretariat shall also communicate proposed amendments to the Signatories to this Convention for information.

3. The Parties shall make every effort to reach agreement on any proposed amendment to this Convention by consensus. If all efforts at consensus have been exhausted, and no agreement reached, the amendment shall as a last resort be adopted by a three-fourths majority vote of the Parties present and voting at the meeting, and shall be submitted by the Depositary to all Parties for ratification, approval, formal confirmation or acceptance.

4. The procedure mentioned in paragraph 3 above shall apply to amendments to any protocol, except that a two-thirds majority of the Parties to that protocol present and voting at the meeting shall suffice for their adoption.

5. Instruments of ratification, approval, formal confirmation or acceptance of amendments shall be deposited with the Depositary. Amendments adopted in accordance with paragraphs 3 or 4 above shall enter into force between Parties having accepted them on the ninetieth day after the receipt by the Depositary of their instrument of ratification, approval, formal confirmation or acceptance by at least three-fourths of the Parties who accepted them or by at least two thirds of the Parties to the protocol concerned who accepted them, except as may otherwise be provided in such protocol. The amendments shall enter into force for any other Party on the ninetieth day after that Party deposits its instrument of ratification, approval, formal confirmation or acceptance of the amendments.

6. For the purpose of this Article, "Parties present and voting" means Parties present and casting an affirmative or negative vote.

ARTICLE 18

ADOPTION AND AMENDMENT OF ANNEXES

1. The annexes to this Convention or to any protocol shall form an integral part of this Convention or of such protocol, as the case may be and, unless expressly provided otherwise, a reference to this Convention or its protocols constitutes at the same time a reference to any annexes thereto. Such annexes shall be restricted to scientific, technical and administrative matters.

2. Except as may be otherwise provided in any protocol with respect to its annexes, the following procedure shall apply to the proposal, adoption and entry into force of additional annexes to this Convention or of annexes to a protocol:

   (a) Annexes to this Convention and its protocols shall be proposed and adopted according to the procedure laid down in Article 17, paragraphs 2, 3 and 4;

   (b) Any Party that is unable to accept an additional annex to this Convention or an annex to any protocol to which it is party shall so notify the Depositary, in writing, within six months from the date of the communication of the adoption by the Depositary. The Depositary shall without delay notify all Parties of any such notification received. A Party may at any time substitute an acceptance for a previous declaration of objection and the annexes shall thereupon enter into force for that Party;

   (c) On the expiry of six months from the date of the circulation of the communication by the Depositary, the annex shall become effective for all Parties to this Convention or to any protocol concerned, which have not submitted a notification in accordance with the provision of subparagraph (b) above.

3. The proposal, adoption and entry into force of amendments to annexes to this Convention or to any protocol shall be subject to the same procedure as for the proposal, adoption and entry into force of annexes to the Convention or annexes to a protocol. Annexes and amendments thereto shall take due account, *inter alia*, of relevant scientific and technical considerations.

4. If an additional annex or an amendment to an annex involves an amendment to this Convention or to any protocol, the additional annex or amended annex shall not enter into force until such time the amendment to this Convention or to the protocol enters into force.

ARTICLE 19

VERIFICATION

Any Party which has reason to believe that another Party is acting or has acted in breach of its obligations under this Convention may inform the Secretariat thereof, and in such an event, shall simultaneously and immediately inform, directly or through the Secretariat, the Party against whom the allegations are made. All relevant information should be submitted by the Secretariat to the Parties.

ARTICLE 20

SETTLEMENT OF DISPUTES

1. In case of a dispute between Parties as to the interpretation or application of, or compliance with, this Convention or any protocol thereto, they shall seek a settlement of the dispute through negotiation or any other peaceful means of their own choice.

2. If the Parties concerned cannot settle their dispute through the means mentioned in the preceding paragraph, the dispute, if the Parties to the dispute agree, shall be submitted to the International Court of Justice or to arbitration under the conditions set out in Annex VI on Arbitration. However, failure to reach common agreement on submission of the dispute to the International Court of Justice or to arbitration shall not absolve the Parties from the responsibility of continuing to seek to resolve it by the means referred to in paragraph 1.

3. When ratifying, accepting, approving, formally confirming or acceding to this Convention, or at any time thereafter, a State or political and/or economic integration organization may declare that it recognizes as compulsory *ipso facto* and without special agreement, in relation to any Party accepting the same obligation:

    (a) submission of the dispute to the International Court of Justice; and/or

    (b) arbitration in accordance with the procedures set out in Annex VI.
    Such declaration shall be notified in writing to the Secretariat which shall communicate it to the Parties.

ARTICLE 21

SIGNATURE

This Convention shall be open for signature by States, by Namibia, represented by the United Nations Council for Namibia, and by political and/or economic integration organizations, in Basel on 22 March 1989, at the Federal Department of Foreign Affairs of Switzerland in Berne from 23 March 1989 to 30 June 1989 and at United Nations Headquarters in New York from 1 July 1989 to 22 March 1990.

ARTICLE 22

RATIFICATION, ACCEPTANCE, FORMAL CONFIRMATION OR APPROVAL

1. This Convention shall be subject to ratification, acceptance or approval by States and by Namibia, represented by the United Nations Council for

Namibia, and to formal confirmation or approval by political and/or economic integration organizations. Instruments of ratification, acceptance, formal confirmation, or approval shall be deposited with the Depositary.

2. Any organization referred to in paragraph 1 above which becomes a Party to this Convention without any of its member States being a Party shall be bound by all the obligations under the Convention. In the case of such organizations, one or more of whose member States is a Party to the Convention, the organization and its member States shall decide on their respective responsibilities for the performance of their obligations under the Convention. In such cases, the organization and the member States shall not be entitled to exercise rights under the Convention concurrently.

3. In their instruments of formal confirmation or approval, the organizations referred to in paragraph 1 above shall declare the extent of their competence with respect to the matters governed by the Convention. These organizations shall also inform the Depositary, who will inform the Parties of any substantial modification in the extent of their competence.

ARTICLE 23

ACCESSION

1. This Convention shall be open for accession by States, by Namibia, represented by the United Nations Council for Namibia, and by political and/or economic integration organizations from the day after the date on which the Convention is closed for signature. The instruments of accession shall be deposited with the Depositary.

2. In their instruments of accession, the organizations referred to in paragraph 1 above shall declare the extent of their competence with respect to the matters governed by the Convention. These organizations shall also inform the Depositary of any substantial modification in the extent of their competence.

3. The provisions of Article 22, paragraph 2, shall apply to political and/or economic integration organizations which accede to this Convention.

ARTICLE 24

RIGHT TO VOTE

1. Except as provided for in paragraph 2 below, each Contracting Party to this Convention shall have one vote.

2. Political and/or economic integration organizations, in matters within their competence, in accordance with Article 22, paragraph 3, and Article 23, paragraph 2, shall exercise their right to vote with a number of votes equal to the number of their member States which are Parties to the Convention or the relevant protocol. Such organizations shall not exercise their right to vote if their member States exercise theirs, and vice versa.

ARTICLE 25

ENTRY INTO FORCE

1. This Convention shall enter into force on the ninetieth day after the date of deposit of the twentieth instrument of ratification, acceptance, formal confirmation, approval or accession.

2. For each State or political and/or economic integration organization which ratifies, accepts, approves or formally confirms this Convention or accedes thereto after the date of the deposit of the twentieth instrument of ratification, acceptance, approval, formal confirmation or accession, it shall enter into force on the ninetieth day after the date of deposit by such State or political and/or economic integration organization of its instrument of ratification, acceptance, approval, formal confirmation or accession.

3. For the purpose of paragraphs 1 and 2 above, any instrument deposited by a political and/or economic integration organization shall not be counted as additional to those deposited by member States of such organization.

ARTICLE 26

RESERVATIONS AND DECLARATIONS

1. No reservation or exception may be made to this Convention.

2. Paragraph 1 of this Article does not preclude a State or political and/or economic integration organization, when signing, ratifying, accepting, approving, formally confirming or acceding to this Convention, from making declarations or statements, however phrased or named, with a view, *inter alia*, to the harmonization of its laws and regulations with the provisions of this Convention, provided that such declarations or statements do not purport to exclude or to modify the legal effects of the provisions of the Convention in their application to that State.

ARTICLE 27

WITHDRAWAL

1. At any time after three years from the date on which this Convention has entered into force for a Party, that Party may withdraw from the Convention by giving written notification to the Depositary.

2. Withdrawal shall be effective one year from receipt of notification by the Depositary, or on such later date as may be specified in the notification.

ARTICLE 28

DEPOSITORY

The Secretary-General of the United Nations shall be the Depository of this Convention and of any protocol thereto.

ARTICLE 29

AUTHENTIC TEXTS

The original Arabic, Chinese, English, French, Russian and Spanish texts of this Convention are equally authentic.

IN WITNESS WHEREOF the undersigned, being duly authorized to that effect, have signed this Convention.

Done at Basel on the 22 day of March 1989

ANNEX I: CATEGORIES OF WASTES TO BE CONTROLLED

WASTE STREAMS

Y1	Clinical wastes from medical care in hospitals, medical centers and clinics
Y2	Wastes from the production and preparation of pharmaceutical products
Y3	Waste pharmaceuticals, drugs and medicines
Y4	Wastes from the production, formulation and use of biocides and phytopharmaceuticals
Y5	Wastes from the manufacture, formulation and use of wood preserving chemicals
Y6	Wastes from the production, formulation and use of organic solvents
Y7	Wastes from heat treatment and tempering operations containing cyanides
Y8	Waste mineral oils unfit for their originally intended use
Y9	Waste oils/water, hydrocarbons/water mixtures, emulsions
Y10	Waste substances and articles containing or contaminated with polychlorinated biphenyls (PCBs) and/or polychlorinated terphenyls (PCTs) and/or polybrominated biphenyls (PBBs)
Y11	Waste tarry residues arising from refining, distillation and any pyrolytic treatment
Y12	Wastes from production, formulation and use of inks, dyes, pigments, paints, lacquers, varnish

(Continued)

Y13	Wastes from production, formulation and use of resins, latex, plasticizers, glues/adhesives
Y14	Waste chemical substances arising from research and development or teaching activities which are not identified and/or are new and whose effects on man and/or the environment are not known
Y15	Wastes of an explosive nature not subject to other legislation
Y16	Wastes from production, formulation and use of photographic chemicals and processing materials
Y17	Wastes resulting from surface treatment of metals and plastics
Y18	Residues arising from industrial waste disposal operations

WASTES HAVING AS CONSTITUENTS

Y19	Metal carbonyls
Y20	Beryllium; beryllium compounds
Y21	Hexavalent chromium compounds
Y22	Copper compounds
Y23	Zinc compounds
Y24	Arsenic; arsenic compounds
Y25	Selenium; selenium compounds
Y26	Cadmium; cadmium compounds
Y27	Antimony; antimony compounds
Y28	Tellurium; tellurium compounds
Y29	Mercury; mercury compounds
Y30	Thallium; thallium compounds
Y31	Lead; lead compounds
Y32	Inorganic fluorine compounds excluding calcium fluoride
Y33	Inorganic cyanides
Y34	Acidic solutions or acids in solid form
Y35	Basic solutions or bases in solid form

(Continued)

Y36	Asbestos (dust and fibres)
Y37	Organic phosphorus compounds
Y38	Organic cyanides
Y39	Phenols; phenol compounds including chlorophenols
Y40	Ethers
Y41	Halogenated organic solvents
Y42	Organic solvents excluding halogenated solvents
Y43	Any congenor of polychlorinated dibenzo-furan
Y44	Any congenor of polychlorinated dibenzo-p-dioxin
Y45	Organohalogen compounds other than substances referred to in this Annex (e.g., Y39, Y41, Y42, Y43, Y44)

(a) To facilitate the application of this Convention, and subject to paragraphs (b), (c) and (d), wastes listed in Annex VIII are characterized as hazardous pursuant to Article 1, paragraph 1 (a), of this Convention, and wastes listed in Annex IX are not covered by Article 1, paragraph 1 (a), of this Convention.

(b) Designation of a waste on Annex VIII does not preclude, in a particular case, the use of Annex III to demonstrate that a waste is not hazardous pursuant to Article 1, paragraph 1 (a), of this Convention.

(c) Designation of a waste on Annex IX does not preclude, in a particular case, characterization of such a waste as hazardous pursuant to Article 1, paragraph 1 (a), of this Convention if it contains Annex I material to an extent causing it to exhibit an Annex III characteristic.

(d) Annexes VIII and IX do not affect the application of Article 1, paragraph 1 (a), of this Convention for the purpose of characterization of wastes.[3]

---

[3]The amendment whereby paragraphs (a), (b), (c) and (d) were added to the end of Annex I entered into force on 6 November 1998, six months following the issuance of depositary notification C.N.77.1998 of 6 May 1998 (reflecting Decision IV/9, adopted by the Conference of the Parties at its fourth meeting).

ANNEX II: CATEGORIES OF WASTES REQUIRING SPECIAL CONSIDERATION

Y46	Wastes collected from households
Y47	Residues arising from the incineration of household wastes

ANNEX III: LIST OF HAZARDOUS CHARACTERISTICS

UN Class[4]	Code	Characteristics
1	H1	Explosive
		An explosive substance or waste is a solid or liquid substance or waste (or mixture of substances or wastes) which is in itself capable by chemical reaction of producing gas at such a temperature and pressure and at such a speed as to cause damage to the surroundings.
3	H3	Flammable liquids
		The word "flammable" has the same meaning as "inflammable." Flammable liquids are liquids, or mixtures of liquids, or liquids containing solids in solution or suspension (for example, paints, varnishes, lacquers, etc., but not including substances or wastes otherwise classified on account of their dangerous characteristics) which give off a flammable vapour at temperatures of not more than 60.5°C, closed-cup test, or not more than 65.6°C, open-cup test. (Since the results of open-cup tests and of closed-cup tests are not strictly comparable and even individual results by the same test are often variable, regulations varying from the above figures to make allowance for such differences would be within the spirit of this definition.)
4.1	H4.1	Flammable solids
		Solids, or waste solids, other than those classed as explosives, which under conditions encountered in transport are readily combustible, or may cause or contribute to fire through friction.

---

[4]Corresponds to the hazard classification system included in the United Nations Recommendations on the Transport of Dangerous Goods (ST/SG/AC.10/1Rev.5, United Nations, New York, 1988).

4.2	H4.2	Substances or wastes liable to spontaneous combustion
		Substances or wastes which are liable to spontaneous heating under normal conditions encountered in transport, or to heating up on contact with air, and being then liable to catch fire.
4.3	H4.3	Substances or wastes which, in contact with water, emit flammable gases
		Substances or wastes which, by interaction with water, are liable to become spontaneously flammable or to give off flammable gases in dangerous quantities.
5.1	H5.1	Oxidizing
		Substances or wastes which, while in themselves not necessarily combustible, may, generally by yielding oxygen cause, or contribute to, the combustion of other materials.
5.2	H5.2	Organic Peroxides
		Organic substances or wastes which contain the bivalent-o-o-structure are thermally unstable substances which may undergo exothermic self-accelerating decomposition.
6.1	H6.1	Poisonous (Acute)
		Substances or wastes liable either to cause death or serious injury or to harm human health if swallowed or inhaled or by skin contact.
6.2	H6.2	Infectious substances
		Substances or wastes containing viable micro organisms or their toxins which are known or suspected to cause disease in animals or humans.
8	H8	Corrosives
		Substances or wastes which, by chemical action, will cause severe damage when in contact with living tissue, or, in the case of leakage, will materially damage, or even destroy, other goods or the means of transport; they may also cause other hazards.
9	H10	Liberation of toxic gases in contact with air or water
		Substances or wastes which, by interaction with air or water, are liable to give off toxic gases in dangerous quantities.
9	H11	Toxic (Delayed or chronic)

(*Continued*)

(Continued)

			Substances or wastes which, if they are inhaled or ingested or if they penetrate the skin, may involve delayed or chronic effects, including carcinogenicity.
9	H12	Ecotoxic	
			Substances or wastes which if released present or may present immediate or delayed adverse impacts to the environment by means of bioaccumulation and/or toxic effects upon biotic systems.
9	H13	Capable, by any means, after disposal, of yielding another material, e.g., leachate, which possesses any of the characteristics listed above.	

TESTS

The potential hazards posed by certain types of wastes are not yet fully documented; tests to define quantitatively these hazards do not exist. Further research is necessary in order to develop means to characterise potential hazards posed to man and/or the environment by these wastes. Standardized tests have been derived with respect to pure substances and materials. Many countries have developed national tests which can be applied to materials listed in Annex I, in order to decide if these materials exhibit any of the characteristics listed in this Annex.

ANNEX IV: DISPOSAL OPERATIONS

A. OPERATIONS WHICH DO NOT LEAD TO THE POSSIBILITY OF RESOURCE RECOVERY, RECYCLING, RECLAMATION, DIRECT REUSE OR ALTERNATIVE USES

Section A encompasses all such disposal operations which occur in practice.

D1	Deposit into or onto land, (e.g., landfill, etc.)
D2	Land treatment, (e.g., biodegradation of liquid or sludgy discards in soils, etc.)
D3	Deep injection, (e.g., injection of pumpable discards into wells, salt domes of naturally occurring repositories, etc.)
D4	Surface impoundment, (e.g., placement of liquid or sludge discards into pits, ponds or lagoons, etc.)

(Continued)

D5	Specially engineered landfill, (e.g., placement into lined discrete cells which are capped and isolated from one another and the environment, etc.)
D6	Release into a water body except seas/oceans
D7	Release into seas/oceans including sea-bed insertion
D8	Biological treatment not specified elsewhere in this Annex which results in final compounds or mixtures which are discarded by means of any of the operations in Section A
D9	Physico chemical treatment not specified elsewhere in this Annex which results in final compounds or mixtures which are discarded by means of any of the operations in Section A, (e.g., evaporation, drying, calcination, neutralization, precipitation, etc.)
D10	Incineration on land
D11	Incineration at sea
D12	Permanent storage (e.g., emplacement of containers in a mine, etc.)
D13	Blending or mixing prior to submission to any of the operations in Section A
D14	Repackaging prior to submission to any of the operations in Section A
D15	Storage pending any of the operations in Section A

B. Operations Which May Lead to Resource Recovery, Recycling Reclamation, Direct Re-use or Alternative Uses

Section B encompasses all such operations with respect to materials legally defined as or considered to be hazardous wastes and which otherwise would have been destined for operations included in Section A

R1	Use as a fuel (other than in direct incineration) or other means to generate energy
R2	Solvent reclamation/regeneration
R3	Recycling/reclamation of organic substances which are not used as solvents
R4	Recycling/reclamation of metals and metal compounds

(*Continued*)

(Continued)

R5	Recycling/reclamation of other inorganic materials
R6	Regeneration of acids or bases
R7	Recovery of components used for pollution abatement
R8	Recovery of components from catalysts
R9	Used oil re-refining or other reuses of previously used oil
R10	Land treatment resulting in benefit to agriculture or ecological improvement
R11	Uses of residual materials obtained from any of the operations numbered R1–R10
R12	Exchange of wastes for submission to any of the operations numbered R1–R11
R13	Accumulation of material intended for any operation in Section B

ANNEX V: A INFORMATION TO BE PROVIDED ON NOTIFICATION

1. Reason for waste export
2. Exporter of the waste [1]
3. Generator(s) of the waste and site of generation [1]
4. Disposer of the waste and actual site of disposal [1]
5. Intended carrier(s) of the waste or their agents, if known [1]
6. Country of export of the waste
   Competent authority [2]
7. Expected countries of transit
   Competent authority [2]
8. Country of import of the waste
   Competent authority [2]
9. General or single notification
10. Projected date(s) of shipment(s) and period of time over which waste is to be exported and proposed itinerary (including point of entry and exit) [3]
11. Means of transport envisaged (road, rail, sea, air, inland waters)
12. Information relating to insurance [4]
13. Designation and physical description of the waste including Y number and UN number and its composition [5] and information on any special handling requirements including emergency provisions in case of accidents
14. Type of packaging envisaged (e.g., bulk, drummed, tanker)

15. Estimated quantity in weight/volume *[6]*

16. Process by which the waste is generated *[7]*

17. For wastes listed in Annex I, classifications from Annex III: hazardous characteristic, H number, and UN class

18. Method of disposal as per Annex IV

19. Declaration by the generator and exporter that the information is correct

20. Information transmitted (including technical description of the plant) to the exporter or generator from the disposer of the waste upon which the latter has based his assessment that there was no reason to believe that the wastes will not be managed in an environmentally sound manner in accordance with the laws and regulations of the country of import

21. Information concerning the contract between the exporter and disposer.

Notes

[1] Full name and address, telephone, telex or telefax number and the name, address, telephone, telex or telefax number of the person to be contacted.

[2] Full name and address, telephone, telex or telefax number.

[3] In the case of a general notification covering several shipments, either the expected dates of each shipment or, if this is not known, the expected frequency of the shipments will be required.

[4] Information to be provided on relevant insurance requirements and how they are met by exporter, carrier and disposer.

[5] The nature and the concentration of the most hazardous components, in terms of toxicity and other dangers presented by the waste both in handling and in relation to the proposed disposal method.

[6] In the case of a general notification covering several shipments, both the estimated total quantity and the estimated quantities for each individual shipment will be required.

[7] Insofar as this is necessary to assess the hazard and determine the appropriateness of the proposed disposal operation.

ANNEX V B: INFORMATION TO BE PROVIDED ON THE MOVEMENT DOCUMENT

1. Exporter of the waste *[1]*

2. Generator(s) of the waste and site of generation *[1]*

3. Disposer of the waste and actual site of disposal *[1]*

4. Carrier(s) of the waste *[1]* or his agent(s)

5. Subject of general or single notification

6. The date the transboundary movement started and date(s) and signature on receipt by each person who takes charge of the waste

7. Means of transport (road, rail, inland waterway, sea, air) including countries of export, transit and import, also point of entry and exit where these have been designated

8. General description of the waste (physical state, proper UN shipping name and class, UN number, Y number and H number as applicable)

9. Information on special handling requirements including emergency provision in case of accidents

10. Type and number of packages

11. Quantity in weight/volume

12. Declaration by the generator or exporter that the information is correct

13. Declaration by the generator or exporter indicating no objection from the competent authorities of all States concerned which are Parties

14. Certification by disposer of receipt at designated disposal facility and indication of method of disposal and of the approximate date of disposal.

Notes

The information required on the movement document shall where possible be integrated in one document with that required under transport rules. Where this is not possible the information should complement rather than duplicate that required under the transport rules. The movement document shall carry instructions as to who is to provide information and fill-out any form.

[1] Full name and address, telephone, telex or telefax number and the name, address, telephone, telex or telefax number of the person to be contacted in case of emergency.

ANNEX VI: ARBITRATION

Article 1

Unless the agreement referred to in Article 20 of the Convention provides otherwise, the arbitration procedure shall be conducted in accordance with Articles 2 to 10 below.

Article 2

The claimant Party shall notify the Secretariat that the Parties have agreed to submit the dispute to arbitration pursuant to paragraph 2 or paragraph 3 of Article 20 and include, in particular, the Articles of the Convention the interpretation or application of which are at issue. The Secretariat shall forward the information thus received to all Parties to the Convention.

Article 3

The arbitral tribunal shall consist of three members. Each of the Parties to the dispute shall appoint an arbitrator, and the two arbitrators so appointed shall designate by common agreement the third arbitrator, who shall be the chairman of the tribunal. The latter shall not be a national of one of the Parties to the dispute, nor have his usual place of residence in the territory of one of these Parties, nor be employed by any of them, nor have dealt with the case in any other capacity.

Article 4

1. If the chairman of the arbitral tribunal has not been designated within two months of the appointment of the second arbitrator, the Secretary-General of the United Nations shall, at the request of either Party, designate him within a further two months period.

2. If one of the Parties to the dispute does not appoint an arbitrator within two months of the receipt of the request, the other Party may inform the Secretary-General of the United Nations who shall designate the chairman of the arbitral tribunal within a further two months' period. Upon designation, the chairman of the arbitral tribunal shall request the Party which has not appointed an arbitrator to do so within two months. After such period, he shall inform the Secretary-General of the United Nations, who shall make this appointment within a further two months' period.

Article 5

1. The arbitral tribunal shall render its decision in accordance with international law and in accordance with the provisions of this Convention.

2. Any arbitral tribunal constituted under the provisions of this Annex shall draw up its own rules of procedure.

Article 6

1. The decisions of the arbitral tribunal both on procedure and on substance, shall be taken by majority vote of its members.

2. The tribunal may take all appropriate measures in order to establish the facts. It may, at the request of one of the Parties, recommend essential interim measures of protection.

3. The Parties to the dispute shall provide all facilities necessary for the effective conduct of the proceedings.

4. The absence or default of a Party in the dispute shall not constitute an impediment to the proceedings.

Article 7

The tribunal may hear and determine counter-claims arising directly out of the subject-matter of the dispute.

Article 8

Unless the arbitral tribunal determines otherwise because of the particular circumstances of the case, the expenses of the tribunal, including the remuneration of its members, shall be borne by the Parties to the dispute in equal shares. The tribunal shall keep a record of all its expenses, and shall furnish a final statement thereof to the Parties.

Article 9

Any Party that has an interest of a legal nature in the subject-matter of the dispute which may be affected by the decision in the case, may intervene in the proceedings with the consent of the tribunal.

Article 10

1. The tribunal shall render its award within five months of the date on which it is established unless it finds it necessary to extend the time-limit for a period which should not exceed five months.
2. The award of the arbitral tribunal shall be accompanied by a statement of reasons. It shall be final and binding upon the Parties to the dispute.
3. Any dispute which may arise between the Parties concerning the interpretation or execution of the award may be submitted by either Party to the arbitral tribunal which made the award or, if the latter cannot be seized thereof, to another tribunal constituted for this purpose in the same manner as the first.

ANNEX VII

[Not yet entered into force.][5]

---

[5] Annex VII is an integral part of the Amendment adopted by the third meeting of the Conference of the Parties in 1995 in its Decision III/1. The amendment is not yet in force. The relevant part of Decision III/1 provides as follows:

"*The Conference*,

. . .

3. *Decides* to adopt the following amendment to the Convention:

'**Annex VII**
Parties and other States which are members of OECD, EC, Liechtenstein.' "

ANNEX VIII: LIST A[6]

Wastes contained in this Annex are characterized as hazardous under Article 1, paragraph 1 (a), of this Convention, and their designation on this Annex does not preclude the use of Annex III to demonstrate that a waste is not hazardous.

A1 Metal and metal-bearing wastes

A1010	Metal wastes and waste consisting of alloys of any of the following:

- Antimony
- Arsenic
- Beryllium
- Cadmium
- Lead
- Mercury
- Selenium
- Tellurium
- Thallium

but excluding such wastes specifically listed on list B.

A1020	Waste having as constituents or contaminants, excluding metal waste in massive form, any of the following:

- Antimony; antimony compounds
- Beryllium; beryllium compounds
- Cadmium; cadmium compounds
- Lead; lead compounds
- Selenium; selenium compounds
- Tellurium; tellurium compounds

(Continued)

---

[6]The amendment whereby Annex VIII was added to the Convention entered into force on 6 November 1998, six months following the issuance of depositary notification C.N.77.1998 of 6 May 1998 (reflecting Decision IV/9 adopted by the Conference of the Parties at its fourth meeting). The amendment to Annex VIII whereby new entries were added entered into force on 20 November 2003 (depositary notification C.N.1314.2003), six months following the issuance of depositary notification C.N.399.2003 of 20 May 2003 (reflecting Decision VI/35 adopted by the Conference of the Parties at its sixth meeting). The amendment to Annex VIII whereby one new entry was added entered into force on 8 October 2005 (depositary notification C.N. 1044.2005), six months following the issuance of depositary notification C.N.263.2005 of 8 April 2005 (re-issued on 13 June 2005, reflecting Decision VII/19 adopted by the Conference of the Parties at its seventh meeting). The present text includes all amendments.

(Continued)

A1030	Wastes having as constituents or contaminants any of the following:
	• Arsenic; arsenic compounds • Mercury; mercury compounds • Thallium; thallium compounds
A1040	Wastes having as constituents any of the following:
	• Metal carbonyls • Hexavalent chromium compounds
A1050	Galvanic sludges
A1060	Waste liquors from the pickling of metals
A1070	Leaching residues from zinc processing, dust and sludges such as jarosite, hematite, etc.
A1080	Waste zinc residues not included on list B, containing lead and cadmium in concentrations sufficient to exhibit Annex III characteristics
A1090	Ashes from the incineration of insulated copper wire
A1100	Dusts and residues from gas cleaning systems of copper smelters
A1110	Spent electrolytic solutions from copper electrorefining and electrowinning operations
A1120	Waste sludges, excluding anode slimes, from electrolyte purification systems in copper electrorefining and electrowinning operations
A1130	Spent etching solutions containing dissolved copper
A1140	Waste cupric chloride and copper cyanide catalysts
A1150	Precious metal ash from incineration of printed circuit boards not included on list B[7]
A1160	Waste lead-acid batteries, whole or crushed
A1170	Unsorted waste batteries excluding mixtures of only list B batteries. Waste batteries not specified on list B containing Annex I constituents to an extent to render them hazardous
A1180	Waste electrical and electronic assemblies or scrap[8] containing components such as accumulators and other batteries included on

---

[7]Note that mirror entry on list B (B1160) does not specify exceptions.
[8]This entry does not include scrap assemblies from electric power generation.

(Continued)

list A, mercury-switches, glass from cathode-ray tubes and other activated glass and PCB-capacitors, or contaminated with Annex I constituents (e.g., cadmium, mercury, lead, polychlorinated biphenyl) to an extent that they possess any of the characteristics contained in Annex III (note the related entry on list B B1110)[9]

A1190 Waste metal cables coated or insulated with plastics containing or contaminated with coal tar, PCB[10], lead, cadmium, other organohalogen compounds or other Annex I constituents to an extent that they exhibit Annex III characteristics.

A2 Wastes containing principally inorganic constituents, which may contain metals and organic materials

A2010 Glass waste from cathode-ray tubes and other activated glasses

A2020 Waste inorganic fluorine compounds in the form of liquids or sludges but excluding such wastes specified on list B

A2030 Waste catalysts but excluding such wastes specified on list B

A2040 Waste gypsum arising from chemical industry processes, when containing Annex I constituents to the extent that it exhibits an Annex III hazardous characteristic (note the related entry on list B B2080)

A2050 Waste asbestos (dusts and fibres)

A2060 Coal-fired power plant fly-ash containing Annex I substances in concentrations sufficient to exhibit Annex III characteristics (note the related entry on list B B2050)

---

[9]PCBs are at a concentration level of 50 mg/kg or more.
[10]PCBs are at a concentration level of 50 mg/kg or more.

A3 Wastes containing principally organic constituents, which may contain metals and inorganic materials

A3010	Waste from the production or processing of petroleum coke and bitumen
A3020	Waste mineral oils unfit for their originally intended use
A3030	Wastes that contain, consist of or are contaminated with leaded anti-knock compound sludges
A3040	Waste thermal (heat transfer) fluids
A3050	Wastes from production, formulation and use of resins, latex, plasticizers, glues/adhesives excluding such wastes specified on list B (note the related entry on list B B4020)
A3060	Waste nitrocellulose
A3070	Waste phenols, phenol compounds including chlorophenol in the form of liquids or sludges
A3080	Waste ethers not including those specified on list B
A3090	Waste leather dust, ash, sludges and flours when containing hexavalent chromium compounds or biocides (note the related entry on list B B3100)
A3100	Waste paring and other waste of leather or of composition leather not suitable for the manufacture of leather articles containing hexavalent chromium compounds or biocides (note the related entry on list B B3090)
A3110	Fellmongery wastes containing hexavalent chromium compounds or biocides or infectious substances (note the related entry on list B B3110)
A3120	Fluff—light fraction from shredding
A3130	Waste organic phosphorous compounds
A3140	Waste non-halogenated organic solvents but excluding such wastes specified on list B
A3150	Waste halogenated organic solvents
A3160	Waste halogenated or unhalogenated non-aqueous distillation residues arising from organic solvent recovery operations
A3170	Wastes arising from the production of aliphatic halogenated hydrocarbons (such as chloromethane, dichloro-ethane, vinyl chloride, vinylidene chloride, allyl chloride and epichlorhydrin)
A3180	Wastes, substances and articles containing, consisting of or contaminated with polychlorinated biphenyl (PCB), polychlorinated terphenyl (PCT), polychlorinated naphthalene

	(PCN) or polybrominated biphenyl (PBB), or any other polybrominated analogues of these compounds, at a concentration level of 50 mg/kg or more[11]
A3190	Waste tarry residues (excluding asphalt cements) arising from refining, distillation and any pyrolitic treatment of organic materials
A3200	Bituminous material (asphalt waste) from road construction and maintenance, containing tar (note the related entry on list B, B2130)

A4 Wastes which may contain either inorganic or organic constituents

A4010	Wastes from the production, preparation and use of pharmaceutical products but excluding such wastes specified on list B
A4020	Clinical and related wastes; that is wastes arising from medical, nursing, dental, veterinary, or similar practices, and wastes generated in hospitals or other facilities during the investigation or treatment of patients, or research projects
A4030	Wastes from the production, formulation and use of biocides and phytopharmaceuticals, including waste pesticides and herbicides which are off-specification, outdated,[12] or unfit for their originally intended use
A4040	Wastes from the manufacture, formulation and use of wood-preserving chemicals[13]
A4050	Wastes that contain, consist of or are contaminated with any of the following: • Inorganic cyanides, excepting precious-metal-bearing residues in solid form containing traces of inorganic cyanides • Organic cyanides

(Continued)

---

[11]The 50 mg/kg level is considered to be an internationally practical level for all wastes. However, many individual countries have established lower regulatory levels (e.g., 20 mg/kg) for specific wastes.

[12]"Outdated" means unused within the period recommended by the manufacturer.

[13]This entry does not include wood treated with wood preserving chemicals.

(Continued)

A4060	Waste oils/water, hydrocarbons/water mixtures, emulsions
A4070	Wastes from the production, formulation and use of inks, dyes, pigments, paints, lacquers, varnish excluding any such waste specified on list B (note the related entry on list B B4010)
A4080	Wastes of an explosive nature (but excluding such wastes specified on list B)
A4090	Waste acidic or basic solutions, other than those specified in the corresponding entry on list B (note the related entry on list B B2120)
A4100	Wastes from industrial pollution control devices for cleaning of industrial off-gases but excluding such wastes specified on list B
A4110	Wastes that contain, consist of or are contaminated with any of the following: • Any congenor of polychlorinated dibenzo-furan • Any congenor of polychlorinated dibenzo-dioxin
A4120	Wastes that contain, consist of or are contaminated with peroxides
A4130	Waste packages and containers containing Annex I substances in concentrations sufficient to exhibit Annex III hazard characteristics
A4140	Waste consisting of or containing off specification or outdated[14] chemicals corresponding to Annex I categories and exhibiting Annex III hazard characteristics
A4150	Waste chemical substances arising from research and development or teaching activities which are not identified and/or are new and whose effects on human health and/or the environment are not known
A4160	Spent activated carbon not included on list B (note the related entry on list B B2060)

---

[14]"Outdated" means unused within the period recommended by the manufacturer.

ANNEX IX: LIST B[15]

Wastes contained in the Annex will not be wastes covered by Article 1, paragraph 1 (a), of this Convention unless they contain Annex I material to an extent causing them to exhibit an Annex III characteristic.

B1 Metal and metal-bearing wastes

B1010	Metal and metal-alloy wastes in metallic, non-dispersible form:

- Precious metals (gold, silver, the platinum group, but not mercury)
- Iron and steel scrap
- Copper scrap
- Nickel scrap
- Aluminium scrap
- Zinc scrap
- Tin scrap
- Tungsten scrap
- Molybdenum scrap
- Tantalum scrap
- Magnesium scrap
- Cobalt scrap
- Bismuth scrap
- Titanium scrap
- Zirconium scrap
- Manganese scrap
- Germanium scrap
- Vanadium scrap
- Scrap of hafnium, indium, niobium, rhenium and gallium
- Thorium scrap
- Rare earths scrap
- Chromium scrap

(Continued)

---

[15]The amendment whereby Annex IX was added to the Convention entered into force on 6 November 1998, six months following the issuance of depositary notification C.N.77.1998 (reflecting Decision IV/9 adopted by the Conference of the Parties at its fourth meeting). The amendment to Annex IX whereby new entries were added entered into force on 20 November 2003 (depositary notification C.N.1314.2003), six months following the issuance of depositary notification C.N.399.2003 of 20 May 2003 (reflecting Decision VI/35 adopted by the Conference of the Parties at its sixth meeting). The amendment to Annex IX whereby one entry was added entered into force on 8 October 2005 (depositary notification C.N.1044.2005), six months following the issuance of depositary notification C.N.263.2005 of 8 April 2005 (re-issued on 13 June 2005, reflecting Decision VII/19 adopted by the Conference of the Parties at its seventh meeting). The present text includes all amendments.

B1020	Clean, uncontaminated metal scrap, including alloys, in bulk finished form (sheet, plate, beams, rods, etc), of:

- Antimony scrap
- Beryllium scrap
- Cadmium scrap
- Lead scrap (but excluding lead-acid batteries)
- Selenium scrap
- Tellurium scrap

B1030	Refractory metals containing residues
B1031	Molybdenum, tungsten, titanium, tantalum, niobium and rhenium metal and metal alloy wastes in metallic dispersible form (metal powder), excluding such wastes as specified in list A under entry A1050, Galvanic sludges
B1040	Scrap assemblies from electrical power generation not contaminated with lubricating oil, PCB or PCT to an extent to render them hazardous
B1050	Mixed non-ferrous metal, heavy fraction scrap, not containing Annex I materials in concentrations sufficient to exhibit Annex III characteristics[16]
B1060	Waste selenium and tellurium in metallic elemental form including powder
B1070	Waste of copper and copper alloys in dispersible form, unless they contain Annex I constituents to an extent that they exhibit Annex III characteristics
B1080	Zinc ash and residues including zinc alloys residues in dispersible form unless containing Annex I constituents in concentration such as to exhibit Annex III characteristics or exhibiting hazard characteristic H4.3[17]
B1090	Waste batteries conforming to a specification, excluding those made with lead, cadmium or mercury

---

[16]Note that even where low level contamination with Annex I materials initially exists, subsequent processes, including recycling processes, may result in separated fractions containing significantly enhanced concentrations of those Annex I materials.

[17]The status of zinc ash is currently under review and there is a recommendation with the United Nations Conference on Trade and Development (UNCTAD) that zinc ashes should not be dangerous goods.

B1100	Metal-bearing wastes arising from melting, smelting and refining of metals:

        • Hard zinc spelter
        • Zinc-containing drosses:
          - Galvanizing slab zinc top dross (>90% Zn)
          - Galvanizing slab zinc bottom dross (>92% Zn)
          - Zinc die casting dross (>85% Zn)
          - Hot dip galvanizers slab zinc dross (batch)(>92% Zn)
          - Zinc skimmings
        • Aluminium skimmings (or skims) excluding salt slag
        • Slags from copper processing for further processing or refining not containing arsenic, lead or cadmium to an extent that they exhibit Annex III hazard characteristics
        • Wastes of refractory linings, including crucibles, originating from copper smelting
        • Slags from precious metals processing for further refining
        • Tantalum-bearing tin slags with less than 0.5% tin

B1110	Electrical and electronic assemblies:

        • Electronic assemblies consisting only of metals or alloys
        • Waste electrical and electronic assemblies or scrap[18] (including printed circuit boards) not containing components such as accumulators and other batteries included on list A, mercury-switches, glass from cathode-ray tubes and other activated glass and PCB-capacitors, or not contaminated with Annex I constituents (e.g., cadmium, mercury, lead, polychlorinated biphenyl) or from which these have been removed, to an extent that they do not possess any of the characteristics contained in Annex III (note the related entry on list A A1180)
        • Electrical and electronic assemblies (including printed circuit boards, electronic components and wires) destined for direct reuse,[19] and not for recycling or final disposal[20]

B1115	Waste metal cables coated or insulated with plastics, not included in list A1190, excluding those destined for Annex IVA operations or any other disposal operations involving, at any stage, uncontrolled thermal processes, such as open-burning.

(Continued)

---

[18]This entry does not include scrap from electrical power generation.
[19]Reuse can include repair, refurbishment or upgrading, but not major reassembly
[20]In some countries these materials destined for direct re-use are not considered wastes.

B1120	Spent catalysts excluding liquids used as catalysts, containing any of:

Transition metals, excluding waste catalysts (spent catalysts, liquid used catalysts or other catalysts) on list A:	Scandium	Titanium
	Vanadium	Chromium
	Manganese	Iron
	Cobalt	Nickel
	Copper	Zinc
	Yttrium	Zirconium
	Niobium	Molybdenum
	Hafnium	Tantalum
	Tungsten	Rhenium
Lanthanides (rare earth metals):	Lanthanum	Cerium
	Praseodymium	Neody
	Samarium	Europium
	Gadolinium	Terbium
	Dysprosium	Holmium
	Erbium	Thulium
	Ytterbium	Lutetium

B1130	Cleaned spent precious-metal-bearing catalysts
B1140	Precious-metal-bearing residues in solid form which contain traces of inorganic cyanides
B1150	Precious metals and alloy wastes (gold, silver, the platinum group, but not mercury) in a dispersible, non-liquid form with appropriate packaging and labelling
B1160	Precious-metal ash from the incineration of printed circuit boards (note the related entry on list A A1150)
B1170	Precious-metal ash from the incineration of photographic film
B1180	Waste photographic film containing silver halides and metallic silver
B1190	Waste photographic paper containing silver halides and metallic silver
B1200	Granulated slag arising from the manufacture of iron and steel
B1210	Slag arising from the manufacture of iron and steel including slags as a source of TiO_2 and vanadium
B1220	Slag from zinc production, chemically stabilized, having a high iron content (above 20%) and processed according to industrial specifications (e.g., DIN 4301) mainly for construction
B1230	Mill scaling arising from the manufacture of iron and steel
B1240	Copper oxide mill-scale
B1250	Waste end-of-life motor vehicles, containing neither liquids nor other hazardous components

B2 Wastes containing principally inorganic constituents, which may contain metals and organic materials

B2010	Wastes from mining operations in non-dispersible form:

- Natural graphite waste
- Slate waste, whether or not roughly trimmed or merely cut, by sawing or otherwise
- Mica waste
- Leucite, nepheline and nepheline syenite waste
- Feldspar waste
- Fluorspar waste
- Silica wastes in solid form excluding those used in foundry operations

B2020	Glass waste in non-dispersible form:

- Cullet and other waste and scrap of glass except for glass from cathode-ray tubes and other activated glasses

B2030	Ceramic wastes in non-dispersible form:

- Cermet wastes and scrap (metal ceramic composites)
- Ceramic based fibres not elsewhere specified or included

B2040	Other wastes containing principally inorganic constituents:

- Partially refined calcium sulphate produced from flue-gas desulphurization (FGD)
- Waste gypsum wallboard or plasterboard arising from the demolition of buildings
- Slag from copper production, chemically stabilized, having a high iron content (above 20%) and processed according to industrial specifications (e.g., DIN 4301 and DIN 8201) mainly for construction and abrasive applications
- Sulphur in solid form
- Limestone from the production of calcium cyanamide (having a pH less than 9)
- Sodium, potassium, calcium chlorides
- Carborundum (silicon carbide)
- Broken concrete
- Lithium-tantalum and lithium-niobium containing glass scraps

B2050	Coal-fired power plant fly-ash, not included on list A (note the related entry on list A A2060)
B2060	Spent activated carbon not containing any Annex I constituents to an extent they exhibit Annex III characteristics, for example, carbon resulting from the treatment of potable water and processes of the food industry and vitamin production (note the related entry on list A, A4160)

(Continued)

(Continued)

B2070	Calcium fluoride sludge
B2080	Waste gypsum arising from chemical industry processes not included on list A (note the related entry on list A A2040)
B2090	Waste anode butts from steel or aluminium production made of petroleum coke or bitumen and cleaned to normal industry specifications (excluding anode butts from chlor alkali electrolyses and from metallurgical industry)
B2100	Waste hydrates of aluminium and waste alumina and residues from alumina production excluding such materials used for gas cleaning, flocculation or filtration processes
B2110	Bauxite residue ("red mud") (pH moderated to less than 11.5)
B2120	Waste acidic or basic solutions with a pH greater than 2 and less than 11.5, which are not corrosive or otherwise hazardous (note the related entry on list A A4090)
B2130	Bituminous material (asphalt waste) from road construction and maintenance, not containing tar[21] (note the related entry on list A, A3200)

---

[21]The concentration level of Benzol (a) pyrene should not be 50 mg/kg or more.

B3 Wastes containing principally organic constituents, which may contain metals and inorganic materials

B3010   Solid plastic waste:
The following plastic or mixed plastic materials, provided they are not mixed with other wastes and are prepared to a specification:

- Scrap plastic of non-halogenated polymers and co-polymers, including but not limited to the following[22]
  - ethylene
  - styrene
  - polypropylene
  - polyethylene terephthalate
  - acrylonitrile
  - butadiene
  - polyacetals
  - polyamides
  - polybutylene terephthalate
  - polycarbonates
  - polyethers
  - polyphenylene sulphides
  - acrylic polymers
  - alkanes C10-C13 (plasticiser)
  - polyurethane (not containing CFCs)
  - polysiloxanes
  - polymethyl methacrylate
  - polyvinyl alcohol
  - polyvinyl butyral
  - polyvinyl acetate
- Cured waste resins or condensation products including the following:
  - urea formaldehyde resins
  - phenol formaldehyde resins
  - melamine formaldehyde resins
  - epoxy resins
  - alkyd resins
  - polyamides
- The following fluorinated polymer wastes[23]
  - perfluoroethylene/propylene (FEP)
  - perfluoro alkoxyl alkane

(Continued)

---

[22]It is understood that such scraps are completely polymerized.

[23]Post-consumer wastes are excluded from this entry:

- Wastes shall not be mixed

- Problems arising from open-burning practices to be considered

- tetrafluoroethylene/per fluoro vinyl ether (PFA)
- tetrafluoroethylene/per fluoro methylvinyl ether (MFA)
- polyvinylfluoride (PVF)
- polyvinylidenefluoride (PVDF)

B3020   Paper, paperboard and paper product wastes
The following materials, provided they are not mixed with
hazardous wastes:
Waste and scrap of paper or paperboard of:

- unbleached paper or paperboard or of corrugated paper or
  paperboard
- other paper or paperboard, made mainly of bleached chemical
  pulp, not coloured in the mass
- paper or paperboard made mainly of mechanical pulp (for
  example, newspapers, journals and similar printed matter)
- other, including but not limited to (1) laminated paperboard (2)
  unsorted scrap

B3030   Textile wastes
The following materials, provided they are not mixed with other
wastes and are prepared to a specification:

- Silk waste (including cocoons unsuitable for reeling, yarn waste
  and garnetted stock)
  - not carded or combed
  - other
- Waste of wool or of fine or coarse animal hair, including yarn
  waste but excluding garnetted stock
  - noils of wool or of fine animal hair
  - other waste of wool or of fine animal hair
  - waste of coarse animal hair
- Cotton waste (including yarn waste and garnetted stock)
  - yarn waste (including thread waste)
  - garnetted stock
  - other
- Flax tow and waste
- Tow and waste (including yarn waste and garnetted stock) of
  true hemp (*Cannabis sativa* L.)
- Tow and waste (including yarn waste and garnetted stock) of jute
  and other textile bast fibres (excluding flax, true hemp and ramie)
- Tow and waste (including yarn waste and garnetted stock) of
  sisal and other textile fibres of the genus Agave
- Tow, noils and waste (including yarn waste and garnetted stock)
  of coconut
- Tow, noils and waste (including yarn waste and garnetted stock)
  of abaca (Manila hemp or *Musa textilis* Nee)

- Tow, noils and waste (including yarn waste and garnetted stock) of ramie and other vegetable textile fibres, not elsewhere specified or included
- Waste (including noils, yarn waste and garnetted stock) of man-made fibres
  - of synthetic fibres
  - of artificial fibres
- Worn clothing and other worn textile articles
- Used rags, scrap twine, cordage, rope and cables and worn out articles of twine, cordage, rope or cables of textile materials
  - sorted
  - other

B3035   Waste textile floor coverings, carpets

B3040   Rubber wastes
The following materials, provided they are not mixed with other wastes:

- Waste and scrap of hard rubber (e.g., ebonite)
- Other rubber wastes (excluding such wastes specified elsewhere)

B3050   Untreated cork and wood waste:

- Wood waste and scrap, whether or not agglomerated in logs, briquettes, pellets or similar forms
- Cork waste: crushed, granulated or ground cork

B3060   Wastes arising from agro-food industries provided it is not infectious:

- Wine lees
- Dried and sterilized vegetable waste, residues and byproducts, whether or not in the form of pellets, of a kind used in animal feeding, not elsewhere specified or included
- Degras: residues resulting from the treatment of fatty substances or animal or vegetable waxes
- Waste of bones and horn-cores, unworked, defatted, simply prepared (but not cut to shape), treated with acid or degelatinised
- Fish waste
- Cocoa shells, husks, skins and other cocoa waste
- Other wastes from the agro-food industry excluding by-products which meet national and international requirements and standards for human or animal consumption

B3065   Waste edible fats and oils of animal or vegetable origin (e.g., frying oils), provided they do not exhibit an Annex III characteristic

(*Continued*)

B3070	The following wastes:
	• Waste of human hair
	• Waste straw
	• Deactivated fungus mycelium from penicillin production to be used as animal feed
B3080	Waste parings and scrap of rubber
B3090	Paring and other wastes of leather or of composition leather not suitable for the manufacture of leather articles, excluding leather sludges, not containing hexavalent chromium compounds and biocides (note the related entry on list A A3100)
B3100	Leather dust, ash, sludges or flours not containing hexavalent chromium compounds or biocides (note the related entry on list A A3090)
B3110	Fellmongery wastes not containing hexavalent chromium compounds or biocides or infectious substances (note the related entry on list A A3110)
B3120	Wastes consisting of food dyes
B3130	Waste polymer ethers and waste non-hazardous monomer ethers incapable of forming peroxides
B3140	Waste pneumatic tyres, excluding those destined for Annex IVA operations

B4 Wastes which may contain either inorganic or organic constituents

B4010	Wastes consisting mainly of water-based/latex paints, inks and hardened varnishes not containing organic solvents, heavy metals or biocides to an extent to render them hazardous (note the related entry on list A A4070)
B4020	Wastes from production, formulation and use of resins, latex, plasticizers, glues/adhesives, not listed on list A, free of solvents and other contaminants to an extent that they do not exhibit Annex III characteristics, e.g., water-based, or glues based on casein starch, dextrin, cellulose ethers, polyvinyl alcohols (note the related entry on list A A3050)
B4030	Used single-use cameras, with batteries not included on list A

Appendix B

Convention on International Trade in Endangered Species of Wild Fauna and Flora

Signed at Washington, D.C., on 3 March 1973
Amended at Bonn, on 22 June 1979

The Contracting States,

Recognizing that wild fauna and flora in their many beautiful and varied forms are an irreplaceable part of the natural systems of the earth which must be protected for this and the generations to come;

Conscious of the ever-growing value of wild fauna and flora from aesthetic, scientific, cultural, recreational and economic points of view;

Recognizing that peoples and States are and should be the best protectors of their own wild fauna and flora;

Recognizing, in addition, that international co-operation is essential for the protection of certain species of wild fauna and flora against over-exploitation through international trade;

Convinced of the urgency of taking appropriate measures to this end; Have agreed as follows:

ARTICLE I

DEFINITIONS

For the purpose of the present Convention, unless the context otherwise requires:

(a) "Species" means any species, subspecies, or geographically separate population thereof;

(b) "Specimen" means:

(i) any animal or plant, whether alive or dead;

(ii) in the case of an animal: for species included in Appendices I and II, any readily recognizable part or derivative thereof; and for species included in Appendix III, any readily recognizable part or derivative thereof specified in Appendix III in relation to the species; and

(iii) in the case of a plant: for species included in Appendix I, any readily recognizable part or derivative thereof; and for species included in Appendices II and III, any readily recognizable part or derivative thereof specified in Appendices II and III in relation to the species;

(c) "Trade" means export, re-export, import and introduction from the sea;

(d) "Re-export" means export of any specimen that has previously been imported;

(e) "Introduction from the sea" means transportation into a State of specimens of any species which were taken in the marine environment not under the jurisdiction of any State;

(f) "Scientific Authority" means a national scientific authority designated in accordance with Article IX;

(g) "Management Authority" means a national management authority designated in accordance with Article IX;

(h) "Party" means a State for which the present Convention has entered into force.

ARTICLE II
FUNDAMENTAL PRINCIPLES

1. Appendix I shall include all species threatened with extinction which are or may be affected by trade. Trade in specimens of these species must be subject to particularly strict regulation in order not to endanger further their survival and must only be authorized in exceptional circumstances.

2. Appendix II shall include:

    (a) all species which although not necessarily now threatened with extinction may become so unless trade in specimens of such species is subject to strict regulation in order to avoid utilization incompatible with their survival; and

    (b) other species which must be subject to regulation in order that trade in specimens of certain species referred to in sub-paragraph (a) of this paragraph may be brought under effective control.

3. Appendix III shall include all species which any Party identifies as being subject to regulation within its jurisdiction for the purpose of preventing or restricting exploitation, and as needing the co-operation of other Parties in the control of trade.

4. The Parties shall not allow trade in specimens of species included in Appendices I, II and III except in accordance with the provisions of the present Convention.

ARTICLE III

REGULATION OF TRADE IN SPECIMENS OF SPECIES INCLUDED IN APPENDIX I

1. All trade in specimens of species included in Appendix I shall be in accordance with the provisions of this Article.

2. The export of any specimen of a species included in Appendix I shall require the prior grant and presentation of an export permit. An export permit shall only be granted when the following conditions have been met:

    (a) a Scientific Authority of the State of export has advised that such export will not be detrimental to the survival of that species;

    (b) a Management Authority of the State of export is satisfied that the specimen was not obtained in contravention of the laws of that State for the protection of fauna and flora;

    (c) a Management Authority of the State of export is satisfied that any living specimen will be so prepared and shipped as to minimize the risk of injury, damage to health or cruel treatment; and

    (d) a Management Authority of the State of export is satisfied that an import permit has been granted for the specimen.

3. The import of any specimen of a species included in Appendix I shall require the prior grant and presentation of an import permit and either an export permit or a re-export certificate. An import permit shall only be granted when the following conditions have been met:

    (a) a Scientific Authority of the State of import has advised that the import will be for purposes which are not detrimental to the survival of the species involved;

    (b) a Scientific Authority of the State of import is satisfied that the proposed recipient of a living specimen is suitably equipped to house and care for it; and

    (c) a Management Authority of the State of import is satisfied that the specimen is not to be used for primarily commercial purposes.

4. The re-export of any specimen of a species included in Appendix I shall require the prior grant and presentation of a re-export certificate. A re-export certificate shall only be granted when the following conditions have been met:

    (a) a Management Authority of the State of re-export is satisfied that the specimen was imported into that State in accordance with the provisions of the present Convention;

    (b) a Management Authority of the State of re-export is satisfied that any living specimen will be so prepared and shipped as to minimize the risk of injury, damage to health or cruel treatment; and

    (c) a Management Authority of the State of re-export is satisfied that an import permit has been granted for any living specimen.

5. The introduction from the sea of any specimen of a species included in Appendix I shall require the prior grant of a certificate from a Management Authority of the State of introduction. A certificate shall only be granted when the following conditions have been met:

   (a) a Scientific Authority of the State of introduction advises that the introduction will not be detrimental to the survival of the species involved;

   (b) a Management Authority of the State of introduction is satisfied that the proposed recipient of a living specimen is suitably equipped to house and care for it; and

   (c) a Management Authority of the State of introduction is satisfied that the specimen is not to be used for primarily commercial purposes.

ARTICLE IV

REGULATION OF TRADE IN SPECIMENS OF SPECIES INCLUDED IN APPENDIX II

1. All trade in specimens of species included in Appendix II shall be in accordance with the provisions of this Article.

2. The export of any specimen of a species included in Appendix II shall require the prior grant and presentation of an export permit. An export permit shall only be granted when the following conditions have been met:

   (a) a Scientific Authority of the State of export has advised that such export will not be detrimental to the survival of that species;

   (b) a Management Authority of the State of export is satisfied that the specimen was not obtained in contravention of the laws of that State for the protection of fauna and flora; and

   (c) a Management Authority of the State of export is satisfied that any living specimen will be so prepared and shipped as to minimize the risk of injury, damage to health or cruel treatment.

3. A Scientific Authority in each Party shall monitor both the export permits granted by that State for specimens of species included in Appendix II and the actual exports of such specimens. Whenever a Scientific Authority determines that the export of specimens of any such species should be limited in order to maintain that species throughout its range at a level consistent with its role in the ecosystems in which it occurs and well above the level at which that species might become eligible for inclusion in Appendix I, the Scientific Authority shall advise the appropriate Management Authority of suitable measures to be taken to limit the grant of export permits for specimens of that species.

4. The import of any specimen of a species included in Appendix II shall require the prior presentation of either an export permit or a re-export certificate.

5. The re-export of any specimen of a species included in Appendix II shall require the prior grant and presentation of a re-export certificate. A re-export certificate shall only be granted when the following conditions have been met:

(a) a Management Authority of the State of re-export is satisfied that the specimen was imported into that State in accordance with the provisions of the present Convention; and

(b) a Management Authority of the State of re-export is satisfied that any living specimen will be so prepared and shipped as to minimize the risk of injury, damage to health or cruel treatment.

6. The introduction from the sea of any specimen of a species included in Appendix II shall require the prior grant of a certificate from a Management Authority of the State of introduction. A certificate shall only be granted when the following conditions have been met:

(a) a Scientific Authority of the State of introduction advises that the introduction will not be detrimental to the survival of the species involved; and

(b) a Management Authority of the State of introduction is satisfied that any living specimen will be so handled as to minimize the risk of injury, damage to health or cruel treatment.

7. Certificates referred to in paragraph 6 of this Article may be granted on the advice of a Scientific Authority, in consultation with other national scientific authorities or, when appropriate, international scientific authorities, in respect of periods not exceeding one year for total numbers of specimens to be introduced in such periods.

ARTICLE V

REGULATION OF TRADE IN SPECIMENS OF SPECIES INCLUDED IN APPENDIX III

1. All trade in specimens of species included in Appendix III shall be in accordance with the provisions of this Article.

2. The export of any specimen of a species included in Appendix III from any State which has included that species in Appendix III shall require the prior grant and presentation of an export permit. An export permit shall only be granted when the following conditions have been met:

(a) a Management Authority of the State of export is satisfied that the specimen was not obtained in contravention of the laws of that State for the protection of fauna and flora; and

(b) a Management Authority of the State of export is satisfied that any living specimen will be so prepared and shipped as to minimize the risk of injury, damage to health or cruel treatment.

3. The import of any specimen of a species included in Appendix III shall require, except in circumstances to which paragraph 4 of this Article applies, the prior presentation of a certificate of origin and, where the import is from a State which has included that species in Appendix III, an export permit.

4. In the case of re-export, a certificate granted by the Management Authority of the State of re-export that the specimen was processed in that State or is being

re-exported shall be accepted by the State of import as evidence that the provisions of the present Convention have been complied with in respect of the specimen concerned.

ARTICLE VI

Permits and Certificates

1. Permits and certificates granted under the provisions of Articles III, IV, and V shall be in accordance with the provisions of this Article.

2. An export permit shall contain the information specified in the model set forth in Appendix IV, and may only be used for export within a period of six months from the date on which it was granted.

3. Each permit or certificate shall contain the title of the present Convention, the name and any identifying stamp of the Management Authority granting it and a control number assigned by the Management Authority.

4. Any copies of a permit or certificate issued by a Management Authority shall be clearly marked as copies only and no such copy may be used in place of the original, except to the extent endorsed thereon.

5. A separate permit or certificate shall be required for each consignment of specimens.

6. A Management Authority of the State of import of any specimen shall cancel and retain the export permit or re-export certificate and any corresponding import permit presented in respect of the import of that specimen.

7. Where appropriate and feasible a Management Authority may affix a mark upon any specimen to assist in identifying the specimen. For these purposes "mark" means any indelible imprint, lead seal or other suitable means of identifying a specimen, designed in such a way as to render its imitation by unauthorized persons as difficult as possible.

ARTICLE VII

Exemptions and Other Special Provisions Relating to Trade

1. The provisions of Articles III, IV and V shall not apply to the transit or transhipment of specimens through or in the territory of a Party while the specimens remain in Customs control.

2. Where a Management Authority of the State of export or re-export is satisfied that a specimen was acquired before the provisions of the present Convention applied to that specimen, the provisions of Articles III, IV and V shall not apply to that specimen where the Management Authority issues a certificate to that effect.

3. The provisions of Articles III, IV and V shall not apply to specimens that are personal or household effects. This exemption shall not apply where:

(a) in the case of specimens of a species included in Appendix I, they were acquired by the owner outside his State of usual residence, and are being imported into that State; or

(b) in the case of specimens of species included in Appendix II:

  (i) they were acquired by the owner outside his State of usual residence and in a State where removal from the wild occurred;

  (ii) they are being imported into the owner's State of usual residence; and

  (iii) the State where removal from the wild occurred requires the prior grant of export permits before any export of such specimens; unless a Management Authority is satisfied that the specimens were acquired before the provisions of the present Convention applied to such specimens.

4. Specimens of an animal species included in Appendix I bred in captivity for commercial purposes, or of a plant species included in Appendix I artificially propagated for commercial purposes, shall be deemed to be specimens of species included in Appendix II.

5. Where a Management Authority of the State of export is satisfied that any specimen of an animal species was bred in captivity or any specimen of a plant species was artificially propagated, or is a part of such an animal or plant or was derived therefrom, a certificate by that Management Authority to that effect shall be accepted in lieu of any of the permits or certificates required under the provisions of Article III, IV or V.

6. The provisions of Articles III, IV and V shall not apply to the non-commercial loan, donation or exchange between scientists or scientific institutions registered by a Management Authority of their State, of herbarium specimens, other preserved, dried or embedded museum specimens, and live plant material which carry a label issued or approved by a Management Authority.

7. A Management Authority of any State may waive the requirements of Articles III, IV and V and allow the movement without permits or certificates of specimens which form part of a travelling zoo, circus, menagerie, plant exhibition or other travelling exhibition provided that:

  (a) the exporter or importer registers full details of such specimens with that Management Authority;

  (b) the specimens are in either of the categories specified in paragraph 2 or 5 of this Article; and

  (c) the Management Authority is satisfied that any living specimen will be so transported and cared for as to minimize the risk of injury, damage to health or cruel treatment.

ARTICLE VIII

MEASURES TO BE TAKEN BY THE PARTIES

1. The Parties shall take appropriate measures to enforce the provisions of the present Convention and to prohibit trade in specimens in violation thereof. These shall include measures:

(a) to penalize trade in, or possession of, such specimens, or both; and

(b) to provide for the confiscation or return to the State of export of such specimens.

2. In addition to the measures taken under paragraph 1 of this Article, a Party may, when it deems it necessary, provide for any method of internal reimbursement for expenses incurred as a result of the confiscation of a specimen traded in violation of the measures taken in the application of the provisions of the present Convention.

3. As far as possible, the Parties shall ensure that specimens shall pass through any formalities required for trade with a minimum of delay. To facilitate such passage, a Party may designate ports of exit and ports of entry at which specimens must be presented for clearance. The Parties shall ensure further that all living specimens, during any period of transit, holding or shipment, are properly cared for so as to minimize the risk of injury, damage to health or cruel treatment.

4. Where a living specimen is confiscated as a result of measures referred to in paragraph 1 of this Article:

(a) the specimen shall be entrusted to a Management Authority of the State of confiscation;

(b) the Management Authority shall, after consultation with the State of export, return the specimen to that State at the expense of that State, or to a rescue centre or such other place as the Management Authority deems appropriate and consistent with the purposes of the present Convention; and

(c) the Management Authority may obtain the advice of a Scientific Authority, or may, whenever it considers it desirable, consult the Secretariat in order to facilitate the decision under sub-paragraph (b) of this paragraph, including the choice of a rescue centre or other place.

5. A rescue centre as referred to in paragraph 4 of this Article means an institution designated by a Management Authority to look after the welfare of living specimens, particularly those that have been confiscated.

6. Each Party shall maintain records of trade in specimens of species included in Appendices I, II and III which shall cover:

(a) the names and addresses of exporters and importers; and

(b) the number and type of permits and certificates granted; the States with which such trade occurred; the numbers or quantities and types of specimens, names of species as included in Appendices I, II and III and, where applicable, the size and sex of the specimens in question.

7. Each Party shall prepare periodic reports on its implementation of the present Convention and shall transmit to the Secretariat:

(a) an annual report containing a summary of the information specified in sub-paragraph (b) of paragraph 6 of this Article; and

(b) a biennial report on legislative, regulatory and administrative measures taken to enforce the provisions of the present Convention.

8. The information referred to in paragraph 7 of this Article shall be available to the public where this is not inconsistent with the law of the Party concerned.

ARTICLE IX

MANAGEMENT AND SCIENTIFIC AUTHORITIES

1. Each Party shall designate for the purposes of the present Convention:

   (a) one or more Management Authorities competent to grant permits or certificates on behalf of that Party; and

   (b) one or more Scientific Authorities.

2. A State depositing an instrument of ratification, acceptance, approval or accession shall at that time inform the Depositary Government of the name and address of the Management Authority authorized to communicate with other Parties and with the Secretariat.

3. Any changes in the designations or authorizations under the provisions of this Article shall be communicated by the Party concerned to the Secretariat for transmission to all other Parties.

4. Any Management Authority referred to in paragraph 2 of this Article shall, if so requested by the Secretariat or the Management Authority of another Party, communicate to it impression of stamps, seals or other devices used to authenticate permits or certificates.

ARTICLE X

TRADE WITH STATES NOT PARTY TO THE CONVENTION

Where export or re-export is to, or import is from, a State not a Party to the present Convention, comparable documentation issued by the competent authorities in that State which substantially conforms with the requirements of the present Convention for permits and certificates may be accepted in lieu thereof by any Party.

ARTICLE XI

CONFERENCE OF THE PARTIES

1. The Secretariat shall call a meeting of the Conference of the Parties not later than two years after the entry into force of the present Convention.

2. Thereafter the Secretariat shall convene regular meetings at least once every two years, unless the Conference decides otherwise, and extraordinary meetings at any time on the written request of at least one-third of the Parties.

3. At meetings, whether regular or extraordinary, the Parties shall review the implementation of the present Convention and may:

   (a) make such provision as may be necessary to enable the Secretariat to carry out its duties, and adopt financial provisions;

(b) consider and adopt amendments to Appendices I and II in accordance with Article XV;

(c) review the progress made towards the restoration and conservation of the species included in Appendices I, II and III;

(d) receive and consider any reports presented by the Secretariat or by any Party; and

(e) where appropriate, make recommendations for improving the effectiveness of the present Convention.

4. At each regular meeting, the Parties may determine the time and venue of the next regular meeting to be held in accordance with the provisions of paragraph 2 of this Article.

5. At any meeting, the Parties may determine and adopt rules of procedure for the meeting.

6. The United Nations, its Specialized Agencies and the International Atomic Energy Agency, as well as any State not a Party to the present Convention, may be represented at meetings of the Conference by observers, who shall have the right to participate but not to vote.

7. Any body or agency technically qualified in protection, conservation or management of wild fauna and flora, in the following categories, which has informed the Secretariat of its desire to be represented at meetings of the Conference by observers, shall be admitted unless at least one-third of the Parties present object:

(a) international agencies or bodies, either governmental or non-governmental, and national governmental agencies and bodies; and

(b) national non-governmental agencies or bodies which have been approved for this purpose by the State in which they are located. Once admitted, these observers shall have the right to participate but not to vote.

ARTICLE XII

THE SECRETARIAT

1. Upon entry into force of the present Convention, a Secretariat shall be provided by the Executive Director of the United Nations Environment Programme. To the extent and in the manner he considers appropriate, he may be assisted by suitable inter-governmental or non-governmental international or national agencies and bodies technically qualified in protection, conservation and management of wild fauna and flora.

2. The functions of the Secretariat shall be:

(a) to arrange for and service meetings of the Parties;

(b) to perform the functions entrusted to it under the provisions of Articles XV and XVI of the present Convention;

(c) to undertake scientific and technical studies in accordance with programmes authorized by the Conference of the Parties as will contribute to the implementation of the present Convention, including studies concerning standards for appropriate preparation and shipment of living specimens and the means of identifying specimens;

(d) to study the reports of Parties and to request from Parties such further information with respect thereto as it deems necessary to ensure implementation of the present Convention;

(e) to invite the attention of the Parties to any matter pertaining to the aims of the present Convention;

(f) to publish periodically and distribute to the Parties current editions of Appendices I, II and III together with any information which will facilitate identification of specimens of species included in those Appendices;

(g) to prepare annual reports to the Parties on its work and on the implementation of the present Convention and such other reports as meetings of the Parties may request;

(h) to make recommendations for the implementation of the aims and provisions of the present Convention, including the exchange of information of a scientific or technical nature;

(i) to perform any other function as may be entrusted to it by the Parties.

ARTICLE XIII

INTERNATIONAL MEASURES

1. When the Secretariat in the light of information received is satisfied that any species included in Appendix I or II is being affected adversely by trade in specimens of that species or that the provisions of the present Convention are not being effectively implemented, it shall communicate such information to the authorized Management Authority of the Party or Parties concerned.

2. When any Party receives a communication as indicated in paragraph 1 of this Article, it shall, as soon as possible, inform the Secretariat of any relevant facts insofar as its laws permit and, where appropriate, propose remedial action. Where the Party considers that an inquiry is desirable, such inquiry may be carried out by one or more persons expressly authorized by the Party.

3. The information provided by the Party or resulting from any inquiry as specified in paragraph 2 of this Article shall be reviewed by the next Conference of the Parties which may make whatever recommendations it deems appropriate.

ARTICLE XIV

EFFECT ON DOMESTIC LEGISLATION AND INTERNATIONAL CONVENTIONS

1. The provisions of the present Convention shall in no way affect the right of Parties to adopt:

(a) stricter domestic measures regarding the conditions for trade, taking, possession or transport of specimens of species included in Appendices I, II and III, or the complete prohibition thereof; or

(b) domestic measures restricting or prohibiting trade, taking, possession or transport of species not included in Appendix I, II or III.

2. The provisions of the present Convention shall in no way affect the provisions of any domestic measures or the obligations of Parties deriving from any treaty, convention, or international agreement relating to other aspects of trade, taking, possession or transport of specimens which is in force or subsequently may enter into force for any Party including any measure pertaining to the Customs, public health, veterinary or plant quarantine fields.

3. The provisions of the present Convention shall in no way affect the provisions of, or the obligations deriving from, any treaty, convention or international agreement concluded or which may be concluded between States creating a union or regional trade agreement establishing or maintaining a common external Customs control and removing Customs control between the parties thereto insofar as they relate to trade among the States members of that union or agreement.

4. A State party to the present Convention, which is also a party to any other treaty, convention or international agreement which is in force at the time of the coming into force of the present Convention and under the provisions of which protection is afforded to marine species included in Appendix II, shall be relieved of the obligations imposed on it under the provisions of the present Convention with respect to trade in specimens of species included in Appendix II that are taken by ships registered in that State and in accordance with the provisions of such other treaty, convention or international agreement.

5. Notwithstanding the provisions of Articles III, IV and V, any export of a specimen taken in accordance with paragraph 4 of this Article shall only require a certificate from a Management Authority of the State of introduction to the effect that the specimen was taken in accordance with the provisions of the other treaty, convention or international agreement in question.

6. Nothing in the present Convention shall prejudice the codification and development of the law of the sea by the United Nations Conference on the Law of the Sea convened pursuant to Resolution 2750 C (XXV) of the General Assembly of the United Nations nor the present or future claims and legal views of any State concerning the law of the sea and the nature and extent of coastal and flag State jurisdiction.

ARTICLE XV

Amendments to Appendices I and II

1. The following provisions shall apply in relation to amendments to Appendices I and II at meetings of the Conference of the Parties:

(a) Any Party may propose an amendment to Appendix I or II for consideration at the next meeting. The text of the proposed amendment shall be communicated to the Secretariat at least 150 days before the meeting. The Secretariat shall consult the other Parties and interested bodies on the amendment in accordance with the provisions of sub-paragraphs (b) and (c) of paragraph 2 of this Article and shall communicate the response to all Parties not later than 30 days before the meeting.

(b) Amendments shall be adopted by a two-thirds majority of Parties present and voting. For these purposes "Parties present and voting" means Parties present and casting an affirmative or negative vote. Parties abstaining from voting shall not be counted among the two-thirds required for adopting an amendment.

(c) Amendments adopted at a meeting shall enter into force 90 days after that meeting for all Parties except those which make a reservation in accordance with paragraph 3 of this Article.

2. The following provisions shall apply in relation to amendments to Appendices I and II between meetings of the Conference of the Parties:

(a) Any Party may propose an amendment to Appendix I or II for consideration between meetings by the postal procedures set forth in this paragraph.

(b) For marine species, the Secretariat shall, upon receiving the text of the proposed amendment, immediately communicate it to the Parties. It shall also consult inter-governmental bodies having a function in relation to those species especially with a view to obtaining scientific data these bodies may be able to provide and to ensuring co-ordination with any conservation measures enforced by such bodies. The Secretariat shall communicate the views expressed and data provided by these bodies and its own findings and recommendations to the Parties as soon as possible.

(c) For species other than marine species, the Secretariat shall, upon receiving the text of the proposed amendment, immediately communicate it to the Parties, and, as soon as possible thereafter, its own recommendations.

(d) Any Party may, within 60 days of the date on which the Secretariat communicated its recommendations to the Parties under sub-paragraph (b) or (c) of this paragraph, transmit to the Secretariat any comments on the proposed amendment together with any relevant scientific data and information.

(e) The Secretariat shall communicate the replies received together with its own recommendations to the Parties as soon as possible.

(f) If no objection to the proposed amendment is received by the Secretariat within 30 days of the date the replies and recommendations were communicated under the provisions of sub-paragraph (e) of this paragraph, the amendment shall enter into force 90 days later for all Parties except those which make a reservation in accordance with paragraph 3 of this Article.

(g) If an objection by any Party is received by the Secretariat, the proposed amendment shall be submitted to a postal vote in accordance with the provisions of sub-paragraphs (h), (i) and (j) of this paragraph.

(h) The Secretariat shall notify the Parties that notification of objection has been received.

(i) Unless the Secretariat receives the votes for, against or in abstention from at least one-half of the Parties within 60 days of the date of notification under sub-paragraph (h) of this paragraph, the proposed amendment shall be referred to the next meeting of the Conference for further consideration.

(j) Provided that votes are received from one-half of the Parties, the amendment shall be adopted by a two-thirds majority of Parties casting an affirmative or negative vote.

(k) The Secretariat shall notify all Parties of the result of the vote.

(l) If the proposed amendment is adopted it shall enter into force 90 days after the date of the notification by the Secretariat of its acceptance for all Parties except those which make a reservation in accordance with paragraph 3 of this Article.

3   During the period of 90 days provided for by sub-paragraph (c) of paragraph 1 or sub-paragraph (l) of paragraph 2 of this Article any Party may by notification in writing to the Depositary Government make a reservation with respect to the amendment. Until such reservation is withdrawn the Party shall be treated as a State not a Party to the present Convention with respect to trade in the species concerned.

ARTICLE XVI

Appendix III and Amendments there to

1. Any Party may at any time submit to the Secretariat a list of species which it identifies as being subject to regulation within its jurisdiction for the purpose mentioned in paragraph 3 of Article II. Appendix III shall include the names of the Parties submitting the species for inclusion therein, the scientific names of the species so submitted, and any parts or derivatives of the animals or plants concerned that are specified in relation to the species for the purposes of sub-paragraph (b) of Article I.

2. Each list submitted under the provisions of paragraph 1 of this Article shall be communicated to the Parties by the Secretariat as soon as possible after receiving it. The list shall take effect as part of Appendix III 90 days after the date of such communication. At any time after the communication of such list, any Party may by notification in writing to the Depositary Government enter a reservation with respect to any species or any parts or derivatives, and until such reservation is withdrawn, the State shall be treated as a State not a Party to the present Convention with respect to trade in the species or part or derivative concerned.

3. A Party which has submitted a species for inclusion in Appendix III may withdraw it at any time by notification to the Secretariat which shall communicate the withdrawal to all Parties. The withdrawal shall take effect 30 days after the date of such communication.

4. Any Party submitting a list under the provisions of paragraph 1 of this Article shall submit to the Secretariat a copy of all domestic laws and regulations applicable to the protection of such species, together with any interpretations which the Party may deem appropriate or the Secretariat may request. The Party shall, for as long as the species in question is included in Appendix III, submit any amendments of such laws and regulations or any interpretations as they are adopted.

ARTICLE XVII
AMENDMENT OF THE CONVENTION

1. An extraordinary meeting of the Conference of the Parties shall be convened by the Secretariat on the written request of at least one-third of the Parties to consider and adopt amendments to the present Convention. Such amendments shall be adopted by a two-thirds majority of Parties present and voting. For these purposes "Parties present and voting" means Parties present and casting an affirmative or negative vote. Parties abstaining from voting shall not be counted among the two-thirds required for adopting an amendment.

2. The text of any proposed amendment shall be communicated by the Secretariat to all Parties at least 90 days before the meeting.

3. An amendment shall enter into force for the Parties which have accepted it 60 days after two-thirds of the Parties have deposited an instrument of acceptance of the amendment with the Depositary Government. Thereafter, the amendment shall enter into force for any other Party 60 days after that Party deposits its instrument of acceptance of the amendment.

ARTICLE XVIII
RESOLUTION OF DISPUTES

1. Any dispute which may arise between two or more Parties with respect to the interpretation or application of the provisions of the present Convention shall be subject to negotiation between the Parties involved in the dispute.

2. If the dispute can not be resolved in accordance with paragraph 1 of this Article, the Parties may, by mutual consent, submit the dispute to arbitration, in particular that of the Permanent Court of Arbitration at The Hague, and the Parties submitting the dispute shall be bound by the arbitral decision.

ARTICLE XIX
SIGNATURE

The present Convention shall be open for signature at Washington until 30th April 1973 and thereafter at Berne until 31st December 1974.

ARTICLE XX
RATIFICATION, ACCEPTANCE, APPROVAL

The present Convention shall be subject to ratification, acceptance or approval. Instruments of ratification, acceptance or approval shall be deposited with the Government of the Swiss Confederation which shall be the Depositary Government.

ARTICLE XXI

ACCESSION

The present Convention shall be open indefinitely for accession. Instruments of accession shall be deposited with the Depositary Government.

ARTICLE XXII

ENTRY INTO FORCE

1. The present Convention shall enter into force 90 days after the date of deposit of the tenth instrument of ratification, acceptance, approval or accession, with the Depositary Government.

2. For each State which ratifies, accepts or approves the present Convention or accedes thereto after the deposit of the tenth instrument of ratification, acceptance, approval or accession, the present Convention shall enter into force 90 days after the deposit by such State of its instrument of ratification, acceptance, approval or accession.

ARTICLE XXIII

RESERVATIONS

1. The provisions of the present Convention shall not be subject to general reservations. Specific reservations may be entered in accordance with the provisions of this Article and Articles XV and XVI.

2. Any State may, on depositing its instrument of ratification, acceptance, approval or accession, enter a specific reservation with regard to:

   (a) any species included in Appendix I, II or III; or

   (b) any parts or derivatives specified in relation to a species included in Appendix III.

3. Until a Party withdraws its reservation entered under the provisions of this Article, it shall be treated as a State not a Party to the present Convention with respect to trade in the particular species or parts or derivatives specified in such reservation.

ARTICLE XXIV

DENUNCIATION

Any Party may denounce the present Convention by written notification to the Depositary Government at any time. The denunciation shall take effect twelve months after the Depositary Government has received the notification.

ARTICLE XXV
DEPOSITARY

1. The original of the present Convention, in the Chinese, English, French, Russian and Spanish languages, each version being equally authentic, shall be deposited with the Depositary Government, which shall transmit certified copies thereof to all States that have signed it or deposited instruments of accession to it.

2. The Depositary Government shall inform all signatory and acceding States and the Secretariat of signatures, deposit of instruments of ratification, acceptance, approval or accession, entry into force of the present Convention, amendments thereto, entry and withdrawal of reservations and notifications of denunciation.

3. As soon as the present Convention enters into force, a certified copy thereof shall be transmitted by the Depositary Government to the Secretariat of the United Nations for registration and publication in accordance with Article 102 of the Charter of the United Nations.

In witness whereof the undersigned Plenipotentiaries, being duly authorized to that effect, have signed the present Convention.

Done at Washington this third day of March, One Thousand Nine Hundred and Seventy-three.

APPENDICES I, II AND III
VALID FROM 22 MAY 2009

Interpretation at http://www.cites.org/eng/app/ interpret.shtml		Appendices in pdf format at http://www.cites.org/eng/app/ e-appendices.pdf
	Appendices	
I	**II**	**III**
FAUNA (ANIMALS) PHYLUM CHORDATA CLASS MAMMALIA (MAMMALS)		
ARTIODACTYLA		
Antilocapridae Pronghorn		
Antilocapra americana (Only the population of Mexico; no other population is included in the Appendices)		

(*Continued*)

(Continued)

Bovidae Antelopes, cattle, duikers, gazelles, goats, sheep, etc.		
Addax nasomaculatus		
	Ammotragus lervia	
		Antilope cervicapra (Nepal)
	Bison bison athabascae	
Bos gaurus (Excludes the domesticated form, which is referenced as *Bos frontalis*, and is not subject to the provisions of the Convention)		
Bos mutus (Excludes the domesticated form, which is referenced as *Bos grunniens*, and is not subject to the provisions of the Convention)		
Bos sauveli		
		Bubalus arnee (Nepal) (Excludes the domesticated form, which is referenced as *Bubalus bubalis*)
Bubalus depressicornis		
Bubalus mindorensis		
Bubalus quarlesi		
	Budorcas taxicolor	
Capra falconeri		
Capricornis milneedwardsii		
Capricornis rubidus		

(Continued)

Capricornis sumatraensis		
Capricornis thar		
	Cephalophus brookei	
	Cephalophus dorsalis	
Cephalophus jentinki		
	Cephalophus ogilbyi	
	Cephalophus silvicultor	
	Cephalophus zebra	
	Damaliscus pygargus pygargus	
Gazella cuvieri		
		Gazella dorcas (Algeria, Tunisia)
Gazella leptoceros		
Hippotragus niger variani		
	Kobus leche	
Naemorhedus baileyi		
Naemorhedus caudatus		
Naemorhedus goral		
Naemorhedus griseus		
Nanger dama		
Oryx dammah		
Oryx leucoryx		
	Ovis ammon (Except the subspecies included in Appendix I)	

(Continued)

(Continued)

Ovis ammon hodgsonii		
Ovis ammon nigrimontana		
	Ovis canadensis (Only the population of Mexico; no other population is included in the Appendices)	
Ovis orientalis ophion		
	Ovis vignei (Except the subspecies included in Appendix I)	
Ovis vignei vignei		
Pantholops hodgsonii		
	Philantomba monticola	
Pseudoryx nghetinhensis		
Rupicapra pyrenaica ornata		
	Saiga borealis	
	Saiga tatarica	
		Tetracerus quadricornis (Nepal)
Camelidae Guanaco, vicuna		
	Lama glama guanicoe	
Vicugna vicugna (Except the populations of: Argentina [the populations of the Provinces of Jujuy and Catamarca and the semi-captive populations of the Provinces of Jujuy, Salta, Catamarca, La Rioja and		

(Continued)

San Juan]; Bolivia [the whole population]; Chile [population of the Primera Región]; and Peru [the whole population]; which are included in Appendix II)		
	Vicugna vicugna (Only the populations of **Argentina**[1] [the populations of the Provinces of Jujuy and Catamarca and the semi-captive populations of the Provinces of Jujuy, Salta, Catamarca, La Rioja and San Juan]; **Bolivia**[2] [the whole population]; **Chile**[3] [population of the Primera Región]; **Peru**[4] [the whole population]; all other populations are included in Appendix I)	
Cervidae Deer, guemals, muntjacs, pudus		
Axis calamianensis		
Axis kuhlii		
Axis porcinus annamiticus		
Blastocerus dichotomus		
	Cervus elaphus bactrianus	
		Cervus elaphus barbarus (Algeria, Tunisia)
Cervus elaphus hanglu		
Dama dama mesopotamica		
Hippocamelus spp.		
		Mazama temama cerasina (Guatemala)

(Continued)

(Continued)

Muntiacus crinifrons		
Muntiacus vuquangensis		
		Odocoileus virginianus mayensis (Guatemala)
Ozotoceros bezoarticus		
	Pudu mephistophiles	
Pudu puda		
Rucervus duvaucelii		
Rucervus eldii		
Hippopotamidae Hippopotamuses		
	Hexaprotodon liberiensis	
	Hippopotamus amphibius	
Moschidae Musk deer		
Moschus **spp.** (Only the populations of Afghanistan, Bhutan, India, Myanmar, Nepal and Pakistan; all other populations are included in Appendix II)		
	Moschus **spp.** (Except the populations of Afghanistan, Bhutan, India, Myanmar, Nepal and Pakistan, which are included in Appendix I)	
Suidae Babirusa, pygmy hog		
Babyrousa babyrussa		
Babyrousa bolabatuensis		
Babyrousa celebensis		
Babyrousa togeanensis		

(Continued)

Sus salvanius		
Tayassuidae Peccaries		
	Tayassuidae spp. (Except the species included in Appendix I and the populations of *Pecari tajacu* of Mexico and the United States of America, which are not included in the Appendices)	
Catagonus wagneri		
CARNIVORA		
Ailuridae Red panda		
Ailurus fulgens		
Canidae Bush dog, foxes, wolves		
		Canis aureus (India)
Canis lupus (Only the populations of Bhutan, India, Nepal and Pakistan; all other populations are included in Appendix II)		
	Canis lupus (Except the populations of Bhutan, India, Nepal and Pakistan, which are included in Appendix I)	
	Cerdocyon thous	
	Chrysocyon brachyurus	
	Cuon alpinus	
	Lycalopex culpaeus	
	Lycalopex fulvipes	
	Lycalopex griseus	
	Lycalopex gymnocercus	

(*Continued*)

(Continued)

Speothos venaticus		
		Vulpes bengalensis (India)
	Vulpes cana	
		Vulpes vulpes griffithi (India)
		Vulpes vulpes montana (India)
		Vulpes vulpes pusilla (India)
	Vulpes zerda	
Eupleridae Fossa, falanouc, Malagasy civet		
	Cryptoprocta ferox	
	Eupleres goudotii	
	Fossa fossana	
Felidae Cats		
	Felidae spp. (Except the species included in Appendix I. Specimens of the domesticated form are not subject to the provisions of the Convention)	
Acinonyx jubatus (Annual export quotas for live specimens and hunting trophies are granted as follows: Botswana: 5; Namibia: 150; Zimbabwe: 50. The trade in such specimens is subject to the provisions of Article III of the Convention)		
Caracal caracal(Only the population of Asia; all other populations		

(Continued)

are included in Appendix II)		
Catopuma temminckii		
Felis nigripes		
Leopardus geoffroyi		
Leopardus jacobitus		
Leopardus pardalis		
Leopardus tigrinus		
Leopardus wiedii		
Lynx pardinus		
Neofelis nebulosa		
Panthera leo persica		
Panthera onca		
Panthera pardus		
Panthera tigris		
Pardofelis marmorata		
Prionailurus bengalensis bengalensis (Only the populations of Bangladesh, India and Thailand; all other populations are included in Appendix II)		
Prionailurus planiceps		
Prionailurus rubiginosus (Only the population of India; all other populations are included in Appendix II)		

(Continued)

(Continued)

Puma concolor coryi		
Puma concolor costaricensis		
Puma concolor couguar		
Puma yagouaroundi (Only the populations of Central and North America; all other populations are included in Appendix II)		
Uncia uncia		
Herpestidae Mongooses		
		Herpestes edwardsi (India)
		Herpestes fuscus (India)
		Herpestes javanicus auropunctatus (India)
		Herpestes smithii (India)
		Herpestes urva (India)
		Herpestes vitticollis (India)
Hyaenidae Aardwolf		
		Proteles cristata (Botswana)
Mephitidae Hog-nosed skunk		
	Conepatus humboldtii	
Mustelidae Badgers, martens, weasels, etc.		
Lutrinae Otters		
	Lutrinae spp. (Except the species included in Appendix I)	

(Continued)

Aonyx capensis microdon (Only the populations of Cameroon and Nigeria; all other populations are included in Appendix II)		
Enhydra lutris nereis		
Lontra felina		
Lontra longicaudis		
Lontra provocax		
Lutra lutra		
Lutra nippon		
Pteronura brasiliensis		
Mustelinae Grisons, honey badger, martens, tayra, weasels		
		Eira barbara (Honduras)
		Galictis vittata (Costa Rica)
		Martes flavigula (India)
		Martes foina intermedia (India)
		Martes gwatkinsii (India)
		Mellivora capensis (Botswana)
		Mustela altaica (India)
		Mustela erminea ferghanae (India)
		Mustela kathiah (India)

(Continued)

(Continued)

Mustela nigripes		
		Mustela sibirica (India)
Odobenidae Walrus		
		Odobenus rosmarus (Canada)
Otariidae Fur seals, sealions		
	Arctocephalus **spp.** (Except the species included in Appendix I)	
Arctocephalus townsendi		
Phocidae Seals		
	Mirounga leonina	
Monachus **spp.**		
Procyonidae Coatis, kinkajou, olingos		
		Bassaricyon gabbii (Costa Rica)
		Bassariscus sumichrasti (Costa Rica)
		Nasua narica (Honduras)
		Nasua nasua solitaria (Uruguay)
		Potos flavus (Honduras)
Ursidae Bears, giant panda		
	Ursidae spp. (Except the species included in Appendix I)	
Ailuropoda melanoleuca		
Helarctos malayanus		
Melursus ursinus		

(Continued)

Tremarctos ornatus		
Ursus arctos (Only the populations of Bhutan, China, Mexico and Mongolia; all other populations are included in Appendix II)		
Ursus arctos isabellinus		
Ursus thibetanus		
Viverridae Binturong, civets, linsangs, otter-civet, palm civets		
		Arctictis binturong (India)
		Civettictis civetta (Botswana)
	Cynogale bennettii	
	Hemigalus derbyanus	
		Paguma larvata (India)
		Paradoxurus hermaphroditus (India)
		Paradoxurus jerdoni (India)
	Prionodon linsang	
Prionodon pardicolor		
		Viverra civettina (India)
		Viverra zibetha (India)
		Viverricula indica (India)

(*Continued*)

(Continued)

CETACEA Dolphins, porpoises, whales		
	CETACEA spp. (Except the species included in Appendix I. A zero annual export quota has been established for live specimens from the Black Sea population of *Tursiops truncatus* removed from the wild and traded for primarily commercial purposes)	
Balaenidae Bowhead whale, right whales		
Balaena mysticetus		
***Eubalaena* spp.**		
Balaenopteridae Humpback whale, rorquals		
Balaenoptera acutorostrata (Except the population of West Greenland, which is included in Appendix II)		
Balaenoptera bonaerensis		
Balaenoptera borealis		
Balaenoptera edeni		
Balaenoptera musculus		
Balaenoptera physalus		
Megaptera novaeangliae		
Delphinidae Dolphins		
Orcaella brevirostris		
***Sotalia* spp.**		
***Sousa* spp.**		
Eschrichtiidae Grey whale		
Eschrichtius robustus		

(Continued)

Iniidae River dolphins		
Lipotes vexillifer		
Neobalaenidae Pygmy right whale		
Caperea marginata		
Phocoenidae Porpoises		
Neophocaena phocaenoides		
Phocoena sinus		
Physeteridae Sperm whales		
Physeter catodon		
Platanistidae River dolphins		
Platanista **spp.**		
Ziphiidae Beaked whales, bottle-nosed whales		
Berardius **spp.**		
Hyperoodon **spp.**		
CHIROPTERA		
Phyllostomidae Broad-nosed bat		
		Platyrrhinus lineatus (Uruguay)
Pteropodidae Fruit bats, flying foxes		
	Acerodon **spp.** (Except the species included in Appendix I)	
Acerodon jubatus		
	Pteropus **spp.** (Except the species included in Appendix I)	
Pteropus insularis		
Pteropus loochoensis		

(Continued)

(Continued)

Pteropus mariannus		
Pteropus molossinus		
Pteropus pelewensis		
Pteropus pilosus		
Pteropus samoensis		
Pteropus tonganus		
Pteropus ualanus		
Pteropus yapensis		
CINGULATA		
Dasypodidae Armadillos		
		Cabassous centralis (Costa Rica)
		Cabassous tatouay (Uruguay)
	Chaetophractus nationi (A zero annual export quota has been established. All specimens shall be deemed to be specimens of species included in Appendix I and the trade in them shall be regulated accordingly)	
Priodontes maximus		
DASYUROMORPHIA		
Dasyuridae Dunnarts		
Sminthopsis longicaudata		
Sminthopsis psammophila		
Thylacinidae Tasmanian wolf, thylacine		
Thylacinus cynocephalus (possibly extinct)		

208

(Continued)

DIPROTODONTIA		
Macropodidae Kangaroos, wallabies		
	Dendrolagus inustus	
	Dendrolagus ursinus	
Lagorchestes hirsutus		
Lagostrophus fasciatus		
Onychogalea fraenata		
Onychogalea lunata		
Phalangeridae Cuscuses		
	Phalanger intercastellanus	
	Phalanger mimicus	
	Phalanger orientalis	
	Spilocuscus kraemeri	
	Spilocuscus maculatus	
	Spilocuscus papuensis	
Potoroidae Rat-kangaroos		
Bettongia **spp.**		
Caloprymnus campestris (possibly extinct)		
Vombatidae Northern hairy-nosed wombat		
Lasiorhinus krefftii		
LAGOMORPHA		
Leporidae Hispid hare, volcano rabbit		
Caprolagus hispidus		
Romerolagus diazi		
MONOTREMATA		

(Continued)

(Continued)

Tachyglossidae Echidnas, spiny anteaters		
	Zaglossus **spp.**	
PERAMELEMORPHIA		
Chaeropodidae Pig-footed bandicoots		
Chaeropus ecaudatus (possibly extinct)		
Peramelidae Bandicoots, echymiperas		
Perameles bougainville		
Thylacomyidae Bilbies		
Macrotis lagotis		
Macrotis leucura		
PERISSODACTYLA		
Equidae Horses, wild asses, zebras		
Equus africanus (Excludes the domesticated form, which is referenced as *Equus asinus*, and is not subject to the provisions of the Convention)		
Equus grevyi		
	Equus hemionus (Except the subspecies included in Appendix I)	
Equus hemionus hemionus		
Equus hemionus khur		
	Equus kiang	
Equus przewalskii		
	Equus zebra hartmannae	

(Continued)

Equus zebra zebra		
Rhinocerotidae Rhinoceroses		
Rhinocerotidae spp. (Except the subspecies included in Appendix II)		
	Ceratotherium simum simum (Only the populations of South Africa and Swaziland; all other populations are included in Appendix I. For the exclusive purpose of allowing international trade in live animals to appropriate and acceptable destinations and hunting trophies. All other specimens shall be deemed to be specimens of species included in Appendix I and the trade in them shall be regulated accordingly)	
Tapiridae Tapirs		
Tapiridae spp. (Except the species included in Appendix II)		
	Tapirus terrestris	
PHOLIDOTA		
Manidae Pangolins		
	Manis **spp.** (A zero annual export quota has been established for *Manis crassicaudata*, *M. culionensis*, *M. javanica* and *M. pentadactyla* for specimens removed from the wild and traded for primarily commercial purposes)	
PILOSA		

(*Continued*)

(Continued)

Bradypodidae Three-toed sloth		
	Bradypus variegatus	
Megalonychidae Two-toed sloth		
		Choloepus hoffmanni (Costa Rica)
Myrmecophagidae American anteaters		
	Myrmecophaga tridactyla	
		Tamandua mexicana (Guatemala)
PRIMATES Apes, monkeys		
	PRIMATES spp. (Except the species included in Appendix I)	
Atelidae Howler and prehensile-tailed monkeys		
Alouatta coibensis		
Alouatta palliata		
Alouatta pigra		
Ateles geoffroyi frontatus		
Ateles geoffroyi panamensis		
Brachyteles arachnoides		
Brachyteles hypoxanthus		
Oreonax flavicauda		
Cebidae New World monkeys		
Callimico goeldii		
Callithrix aurita		
Callithrix flaviceps		
Leontopithecus **spp.**		

(Continued)

Saguinus bicolor		
Saguinus geoffroyi		
Saguinus leucopus		
Saguinus martinsi		
Saguinus oedipus		
Saimiri oerstedii		
Cercopithecidae Old World monkeys		
Cercocebus galeritus		
Cercopithecus diana		
Cercopithecus roloway		
Macaca silenus		
Mandrillus leucophaeus		
Mandrillus sphinx		
Nasalis larvatus		
Piliocolobus kirkii		
Piliocolobus rufomitratus		
Presbytis potenziani		
Pygathrix spp.		
Rhinopithecus spp.		
Semnopithecus ajax		
Semnopithecus dussumieri		
Semnopithecus entellus		
Semnopithecus hector		
Semnopithecus hypoleucos		
Semnopithecus priam		

(Continued)

(Continued)

Semnopithecus schistaceus		
Simias concolor		
Trachypithecus geei		
Trachypithecus pileatus		
Trachypithecus shortridgei		
Cheirogaleidae Dwarf lemurs		
Cheirogaleidae spp.		
Daubentoniidae Aye-aye		
Daubentonia madagascariensis		
Hominidae Chimpanzees, gorilla, orang-utan		
Gorilla beringei		
Gorilla gorilla		
Pan **spp.**		
Pongo abelii		
Pongo pygmaeus		
Hylobatidae Gibbons		
Hylobatidae spp.		
Indriidae Avahi, indris, sifakas, woolly lemurs		
Indriidae spp.		
Lemuridae Large lemurs		
Lemuridae spp.		
Lepilemuridae Sportive lemurs		
Lepilemuridae spp.		
Lorisidae Lorises		
Nycticebus **spp.**		

(Continued)

Pithecidae Sakis and uakaris		
***Cacajao* spp.**		
Chiropotes albinasus		
PROBOSCIDEA		
Elephantidae Elephants		
Elephas maximus		
Loxodonta africana (Except the populations of Botswana, Namibia, South Africa and Zimbabwe, which are included in Appendix II)		
	***Loxodonta africana*[5]** (Only the populations of **Botswana, Namibia, South Africa** and **Zimbabwe**; all other populations are included in Appendix I)	
RODENTIA		
Chinchillidae Chinchillas		
***Chinchilla* spp.** (Specimens of the domesticated form are not subject to the provisions of the Convention)		
Cuniculidae Paca		
		Cuniculus paca (Honduras)
Dasyproctidae Agouti		
		Dasyprocta punctata (Honduras)
Erethizontidae New World porcupines		

(Continued)

(Continued)

		Sphiggurus mexicanus (Honduras)
		Sphiggurus spinosus (Uruguay)
Muridae Mice, rats		
Leporillus conditor		
Pseudomys fieldi praeconis		
Xeromys myoides		
Zyzomys pedunculatus		
Sciuridae Ground squirrels, tree squirrels		
Cynomys mexicanus		
		Marmota caudata (India)
		Marmota himalayana (India)
	Ratufa **spp.**	
		Sciurus deppei (Costa Rica)
SCANDENTIA Tree shrews		
	SCANDENTIA spp.	
SIRENIA		
Dugongidae Dugong		
Dugong dugon		
Trichechidae Manatees		
Trichechus inunguis		
Trichechus manatus		
	Trichechus senegalensis	
CLASS AVES (BIRDS)		

(Continued)

ANSERIFORMES		
Anatidae Ducks, geese, swans, etc.		
Anas aucklandica		
	Anas bernieri	
Anas chlorotis		
	Anas formosa	
Anas laysanensis		
Anas nesiotis		
Anas oustaleti		
Asarcornis scutulata		
Branta canadensis leucopareia		
	Branta ruficollis	
Branta sandvicensis		
		Cairina moschata (Honduras)
	Coscoroba coscoroba	
	Cygnus melancoryphus	
	Dendrocygna arborea	
		Dendrocygna autumnalis (Honduras)
		Dendrocygna bicolor (Honduras)
	Oxyura leucocephala	
Rhodonessa caryophyllacea (possibly extinct)		
	Sarkidiornis melanotos	
APODIFORMES		

(*Continued*)

(Continued)

Trochilidae Hummingbirds		
	Trochilidae spp. (Except the species included in Appendix I)	
Glaucis dohrnii		
CHARADRIIFORMES		
Burhinidae Thick-knee		
		Burhinus bistriatus (Guatemala)
Laridae Gull		
Larus relictus		
Scolopacidae Curlews, greenshanks		
Numenius borealis		
Numenius tenuirostris		
Tringa guttifer		
CICONIIFORMES		
Balaenicipitidae Shoebill, whale-headed stork		
	Balaeniceps rex	
Ciconiidae Storks		
Ciconia boyciana		
	Ciconia nigra	
Jabiru mycteria		
Mycteria cinerea		
Phoenicopteridae Flamingos		
	Phoenicopteridae spp.	
Threskiornithidae Ibises, spoonbills		
	Eudocimus ruber	
	Geronticus calvus	
Geronticus eremita		

(Continued)

Nipponia nippon		
	Platalea leucorodia	
COLUMBIFORMES		
Columbidae Doves, pigeons		
Caloenas nicobarica		
Ducula mindorensis		
	Gallicolumba luzonica	
	Goura **spp.**	
		Nesoenas mayeri (Mauritius)
CORACIIFORMES		
Bucerotidae Hornbills		
	Aceros **spp.** (Except the species included in Appendix I)	
Aceros nipalensis		
	Anorrhinus **spp.**	
	Anthracoceros **spp.**	
	Berenicornis **spp.**	
	Buceros **spp.** (Except the species included in Appendix I)	
Buceros bicornis		
	Penelopides **spp.**	
Rhinoplax vigil		
	Rhyticeros **spp.** (Except the species included in Appendix I)	
Rhyticeros subruficollis		
CUCULIFORMES		
Musophagidae Turacos		

(*Continued*)

(Continued)

	Tauraco **spp.**	
FALCONIFORMES Eagles, falcons, hawks, vultures		
	FALCONIFORMES spp. (Except the species included in Appendices I and III and the species of the family Cathartidae)	
Accipitridae Hawks, eagles		
Aquila adalberti		
Aquila heliaca		
Chondrohierax uncinatus wilsonii		
Haliaeetus albicilla		
Harpia harpyja		
Pithecophaga jefferyi		
Cathartidae New World vultures		
Gymnogyps californianus		
		Sarcoramphus papa (Honduras)
Vultur gryphus		
Falconidae Falcons		
Falco araeus		
Falco jugger		
Falco newtoni (Only the population of Seychelles)		
Falco pelegrinoides		
Falco peregrinus		
Falco punctatus		
Falco rusticolus		

(Continued)

GALLIFORMES		
Cracidae Chachalacas, currassows, guans		
		Crax alberti (Colombia)
Crax blumenbachii		
		Crax daubentoni (Colombia)
		Crax globulosa (Colombia)
		Crax rubra (Colombia, Costa Rica, Guatemala, Honduras)
Mitu mitu		
Oreophasis derbianus		
		Ortalis vetula (Guatemala, Honduras)
		Pauxi pauxi (Colombia)
Penelope albipennis		
		Penelope purpurascens (Honduras)
		Penelopina nigra (Guatemala)
Pipile jacutinga		
Pipile pipile		
Megapodiidae Megapodes, scrubfowl		
Macrocephalon maleo		
Phasianidae Grouse, guineafowl, partridges, pheasants, tragopans		
	Argusianus argus	

(*Continued*)

(Continued)

Catreus wallichii		
Colinus virginianus ridgwayi		
Crossoptilon crossoptilon		
Crossoptilon mantchuricum		
	Gallus sonneratii	
	Ithaginis cruentus	
Lophophorus impejanus		
Lophophorus lhuysii		
Lophophorus sclateri		
Lophura edwardsi		
Lophura imperialis		
Lophura swinhoii		
		Meleagris ocellata (Guatemala)
	Pavo muticus	
	Polyplectron bicalcaratum	
	Polyplectron germaini	
	Polyplectron malacense	
Polyplectron napoleonis		
	Polyplectron schleiermacheri	
Rheinardia ocellata		
Syrmaticus ellioti		
Syrmaticus humiae		
Syrmaticus mikado		
Tetraogallus caspius		

(Continued)

Tetraogallus tibetanus		
Tragopan blythii		
Tragopan caboti		
Tragopan melanocephalus		
		Tragopan satyra (Nepal)
Tympanuchus cupido attwateri		
GRUIFORMES		
Gruidae Cranes		
	Gruidae spp. (Except the species included in Appendix I)	
Grus americana		
Grus canadensis nesiotes		
Grus canadensis pulla		
Grus japonensis		
Grus leucogeranus		
Grus monacha		
Grus nigricollis		
Grus vipio		
Otididae Bustards		
	Otididae spp. (Except the species included in Appendix I)	
Ardeotis nigriceps		
Chlamydotis macqueenii		
Chlamydotis undulata		

(Continued)

(Continued)

Houbaropsis bengalensis		
Rallidae Rail		
Gallirallus sylvestris		
Rhynochetidae Kagu		
Rhynochetos jubatus		
PASSERIFORMES		
Atrichornithidae Scrub-bird		
Atrichornis clamosus		
Cotingidae Cotingas		
		Cephalopterus ornatus (Colombia)
		Cephalopterus penduliger (Colombia)
Cotinga maculata		
	Rupicola **spp.**	
Xipholena atropurpurea		
Emberizidae Cardinals, tanagers		
	Gubernatrix cristata	
	Paroaria capitata	
	Paroaria coronata	
	Tangara fastuosa	
Estrildidae Mannikins, waxbills		
	Amandava formosa	
	Lonchura oryzivora	
	Poephila cincta cincta	
Fringillidae Finches		
Carduelis cucullata		

(Continued)

	Carduelis yarrellii	
Hirundinidae Martin		
Pseudochelidon sirintarae		
Icteridae Blackbird		
Xanthopsar flavus		
Meliphagidae Honeyeater		
Lichenostomus melanops cassidix		
Muscicapidae Old World flycatchers		
		Acrocephalus rodericanus (Mauritius)
	Cyornis ruckii	
Dasyornis broadbenti litoralis (possibly extinct)		
Dasyornis longirostris		
	Garrulax canorus	
	Leiothrix argentauris	
	Leiothrix lutea	
	Liocichla omeiensis	
Picathartes gymnocephalus		
Picathartes oreas		
		Terpsiphone bourbonnensis (Mauritius)
Paradisaeidae Birds of paradise		
	Paradisaeidae spp.	

(Continued)

(Continued)

Pittidae Pittas		
	Pitta guajana	
Pitta gurneyi		
Pitta kochi		
	Pitta nympha	
Pycnonotidae Bulbul		
	Pycnonotus zeylanicus	
Sturnidae Mynahs (Starlings)		
	Gracula religiosa	
Leucopsar rothschildi		
Zosteropidae White-eye		
Zosterops albogularis		
PELECANIFORMES		
Fregatidae Frigatebird		
Fregata andrewsi		
Pelecanidae Pelican		
Pelecanus crispus		
Sulidae Booby		
Papasula abbotti		
PICIFORMES		
Capitonidae Barbet		
		Semnornis ramphastinus (Colombia)
Picidae Woodpeckers		
Campephilus imperialis		
Dryocopus javensis richardsi		

(Continued)

Ramphastidae Toucans		
		Baillonius bailloni (Argentina)
	Pteroglossus aracari	
		Pteroglossus castanotis (Argentina)
	Pteroglossus viridis	
		Ramphastos dicolorus (Argentina)
	Ramphastos sulfuratus	
	Ramphastos toco	
	Ramphastos tucanus	
	Ramphastos vitellinus	
		Selenidera maculirostris (Argentina)
PODICIPEDIFORMES		
Podicipedidae Grebe		
Podilymbus gigas		
PROCELLARIIFORMES		
Diomedeidae Albatross		
Phoebastria albatrus		
PSITTACIFORMES		
	PSITTACIFORMES spp. (Except the species included in Appendix I and *Agapornis roseicollis, Melopsittacus undulatus, Nymphicus hollandicus* and *Psittacula krameri*, which are not included in the Appendices)	
Cacatuidae Cockatoos		

(Continued)

(Continued)

Cacatua goffini		
Cacatua haematuropygia		
Cacatua moluccensis		
Cacatua sulphurea		
Prosciger aterrimus		
Loriidae Lories, lorikeets		
Eos histrio		
Vini ultramarina		
Psittacidae Amazons, macaws, parakeets, parrots		
Amazona arausiaca		
Amazona auropalliata		
Amazona barbadensis		
Amazona brasiliensis		
Amazona finschi		
Amazona guildingii		
Amazona imperialis		
Amazona leucocephala		
Amazona oratrix		
Amazona pretrei		
Amazona rhodocorytha		
Amazona tucumana		
Amazona versicolor		
Amazona vinacea		
Amazona viridigenalis		
Amazona vittata		
Anodorhynchus spp.		

(Continued)

Ara ambiguus		
Ara glaucogularis (Often traded under the incorrect designation *Ara caninde*)		
Ara macao		
Ara militaris		
Ara rubrogenys		
Cyanopsitta spixii		
Cyanoramphus cookii		
Cyanoramphus forbesi		
Cyanoramphus novaezelandiae		
Cyanoramphus saisseti		
Cyclopsitta diophthalma coxeni		
Eunymphicus cornutus		
Guarouba guarouba		
Neophema chrysogaster		
Ognorhynchus icterotis		
Pezoporus occidentalis (possibly extinct)		
Pezoporus wallicus		
Pionopsitta pileata		
Primolius couloni		
Primolius maracana		

(*Continued*)

(Continued)

Psephotus chrysopterygius		
Psephotus dissimilis		
Psephotus pulcherrimus (possibly extinct)		
Psittacula echo		
Pyrrhura cruentata		
Rhynchopsitta **spp.**		
Strigops habroptilus		
RHEIFORMES		
Rheidae Rheas		
Pterocnemia pennata (Except *Pterocnemia pennata pennata* which is included in Appendix II)		
	Pterocnemia pennata pennata	
	Rhea americana	
SPHENISCIFORMES		
Spheniscidae Penguins		
	Spheniscus demersus	
Spheniscus humboldti		
STRIGIFORMES Owls		
	STRIGIFORMES spp. (Except the species included in Appendix I)	
Strigidae Owls		
Heteroglaux blewitti		
Mimizuku gurneyi		
Ninox natalis		

(Continued)

Ninox novaeseelandiae undulata		
Tytonidae Barn owls		
Tyto soumagnei		
STRUTHIONIFORMES		
Struthionidae Ostrich		
Struthio camelus (Only the populations of Algeria, Burkina Faso, Cameroon, the Central African Republic, Chad, Mali, Mauritania, Morocco, the Niger, Nigeria, Senegal and the Sudan; all other populations are not included in the Appendices)		
TINAMIFORMES		
Tinamidae Tinamous		
Tinamus solitarius		
TROGONIFORMES		
Trogonidae Quetzals		
Pharomachrus mocinno		
CLASS REPTILIA (REPTILES)		
CROCODYLIA Alligators, caimans, crocodiles		
	CROCODYLIA spp. (Except the species included in Appendix I)	
Alligatoridae Alligators, caimans		
Alligator sinensis		

(Continued)

(Continued)

Caiman crocodilus apaporiensis		
Caiman latirostris (Except the population of Argentina, which is included in Appendix II)		
Melanosuchus niger (Except the population of Brazil, which is included in Appendix II, and the population of Ecuador, which is included in Appendix II and is subject to a zero annual export quota until an annual export quota has been approved by the CITES Secretariat and the IUCN/SSC Crocodile Specialist Group)		
Crocodylidae Crocodiles		
Crocodylus acutus (Except the population of Cuba, which is included in Appendix II)		
Crocodylus cataphractus		
Crocodylus intermedius		
Crocodylus mindorensis		
Crocodylus moreletii		
Crocodylus niloticus (Except the populations of Botswana, Ethiopia, Kenya, Madagascar, Malawi, Mozambique,		

(Continued)

Namibia, South Africa, Uganda, the United Republic of Tanzania [subject to an annual export quota of no more than 1,600 wild specimens including hunting trophies, in addition to ranched specimens], Zambia and Zimbabwe, which are included in Appendix II)		
Crocodylus palustris		
Crocodylus porosus (Except the populations of Australia, Indonesia and Papua New Guinea, which are included in Appendix II)		
Crocodylus rhombifer		
Crocodylus siamensis		
Osteolaemus tetraspis		
Tomistoma schlegelii		
Gavialidae Gavial		
Gavialis gangeticus		
RHYNCHOCEPHALIA		
Sphenodontidae Tuatara		
***Sphenodon* spp.**		
SAURIA		
Agamidae Agamas, mastigures		
	***Uromastyx* spp.**	
Chamaeleonidae Chameleons		

(*Continued*)

(Continued)

	Bradypodion spp.	
	Brookesia spp. (Except the species included in Appendix I)	
Brookesia perarmata		
	Calumma spp.	
	Chamaeleo spp.	
	Furcifer spp.	
Cordylidae Spiny-tailed lizards		
	Cordylus spp.	
Gekkonidae Geckos		
	Cyrtodactylus serpensinsula	
		Hoplodactylus spp. (New Zealand)
		Naultinus spp. (New Zealand)
	Phelsuma spp.	
	Uroplatus spp.	
Helodermatidae Beaded lizard, gila monster		
	Heloderma spp. (Except the subspecies included in Appendix I)	
Heloderma horridum charlesbogerti		
Iguanidae Iguanas		
	Amblyrhynchus cristatus	
Brachylophus spp.		
	Conolophus spp.	
Cyclura spp.		
	Iguana spp.	
	Phrynosoma coronatum	

(Continued)

Sauromalus varius		
Lacertidae Lizards		
Gallotia simonyi		
	Podarcis lilfordi	
	Podarcis pityusensis	
Scincidae Skinks		
	Corucia zebrata	
Teiidae Caiman lizards, tegu lizards		
	Crocodilurus amazonicus	
	Dracaena **spp.**	
	Tupinambis **spp.**	
Varanidae Monitor lizards		
	Varanus **spp.** (Except the species included in Appendix I)	
Varanus bengalensis		
Varanus flavescens		
Varanus griseus		
Varanus komodoensis		
Varanus nebulosus		
Xenosauridae Chinese crocodile lizard		
	Shinisaurus crocodilurus	
SERPENTES Snakes		
Boidae Boas		
	Boidae spp. (Except the species included in Appendix I)	
Acrantophis **spp.**		
Boa constrictor occidentalis		

(*Continued*)

(Continued)

Epicrates inornatus		
Epicrates monensis		
Epicrates subflavus		
Sanzinia madagascariensis		
Bolyeriidae Round Island boas		
	Bolyeriidae spp. (Except the species included in Appendix I)	
Bolyeria multocarinata		
Casarea dussumieri		
Colubridae Typical snakes, water snakes, whipsnakes		
		Atretium schistosum (India)
		Cerberus rynchops (India)
	Clelia clelia	
	Cyclagras gigas	
	Elachistodon westermanni	
	Ptyas mucosus	
		Xenochrophis piscator (India)
Elapidae Cobras, coral snakes		
	Hoplocephalus bungaroides	
		Micrurus diastema (Honduras)
		Micrurus nigrocinctus (Honduras)
	Naja atra	
	Naja kaouthia	
	Naja mandalayensis	

(Continued)

	Naja naja	
	Naja oxiana	
	Naja philippinensis	
	Naja sagittifera	
	Naja samarensis	
	Naja siamensis	
	Naja sputatrix	
	Naja sumatrana	
	Ophiophagus hannah	
Loxocemidae Mexican dwarf boa		
	Loxocemidae spp.	
Pythonidae Pythons		
	Pythonidae spp. (Except the subspecies included in Appendix I)	
Python molurus molurus		
Tropidophiidae Wood boas		
	Tropidophiidae spp.	
Viperidae Vipers		
		Crotalus durissus (Honduras)
		Daboia russelii (India)
Vipera ursinii (Only the population of Europe, except the area which formerly constituted the Union of Soviet Socialist Republics; these latter populations are not included in the Appendices)		

(*Continued*)

(Continued)

	Vipera wagneri	
TESTUDINES		
Carettochelyidae Pig-nosed turtles		
	Carettochelys insculpta	
Chelidae Austro-American side-necked turtles		
	Chelodina mccordi	
Pseudemydura umbrina		
Cheloniidae Marine turtles		
Cheloniidae spp.		
Chelydridae Snapping turtles		
		Macrochelys temminckii (United States of America)
Dermatemydidae Central American river turtle		
	Dermatemys mawii	
Dermochelyidae Leatherback turtle		
Dermochelys coriacea		
Emydidae Box turtles, freshwater turtles		
	Glyptemys insculpta	
Glyptemys muhlenbergii		
		Graptemys **spp.** (United States of America)
	Terrapene **spp.** (Except the species included in Appendix I)	
Terrapene coahuila		
Geoemydidae Box turtles, freshwater turtles		
Batagur baska		

(Continued)

	Callagur borneoensis	
	Cuora spp.	
Geoclemys hamiltonii		
		Geoemyda spengleri (China)
	Heosemys annandalii	
	Heosemys depressa	
	Heosemys grandis	
	Heosemys spinosa	
	Kachuga spp.	
	Leucocephalon yuwonoi	
	Malayemys macrocephala	
	Malayemys subtrijuga	
	Mauremys annamensis	
		Mauremys iversoni (China)
		Mauremys megalocephala (China)
	Mauremys mutica	
		Mauremys nigricans (China)
		Mauremys pritchardi (China)
		Mauremys reevesii (China)
		Mauremys sinensis (China)
Melanochelys tricarinata		
Morenia ocellata		

(*Continued*)

(Continued)

	Notochelys platynota	
		Ocadia glyphistoma (China)
		Ocadia philippeni (China)
	Orlitia borneensis	
	Pangshura spp. (Except the species included in Appendix I)	
Pangshura tecta		
		Sacalia bealei (China)
		Sacalia pseudocellata (China)
		Sacalia quadriocellata (China)
	Siebenrockiella crassicollis	
	Siebenrockiella leytensis	
Platysternidae Big-headed turtle		
	Platysternon megacephalum	
Podocnemididae Afro-American side-necked turtles		
	Erymnochelys madagascariensis	
	Peltocephalus dumerilianus	
	Podocnemis spp.	
Testudinidae Tortoises		
	Testudinidae spp. (Except the species included in Appendix I. A zero annual export quota has been established for *Geochelone sulcata* for specimens removed from the wild and traded for primarily commercial purposes)	
Astrochelys radiata		

(Continued)

Astrochelys yniphora		
Chelonoidis nigra		
Gopherus flavomarginatus		
Psammobates geometricus		
Pyxis arachnoides		
Pyxis planicauda		
Testudo kleinmanni		
Trionychidae Softshell turtles, terrapins		
	Amyda cartilaginea	
Apalone spinifera atra		
Aspideretes gangeticus		
Aspideretes hurum		
Aspideretes nigricans		
	Chitra **spp.**	
	Lissemys punctata	
	Lissemys scutata	
		Palea steindachneri (China)
	Pelochelys **spp.**	
		Pelodiscus axenaria (China)
		Pelodiscus maackii (China)
		Pelodiscus parviformis (China)
		Rafetus swinhoei (China)

(*Continued*)

(Continued)

CLASS AMPHIBIA (AMPHIBIANS)		
ANURA		
Bufonidae Toads		
Altiphrynoides spp.		
Atelopus zeteki		
Bufo periglenes		
Bufo superciliaris		
Nectophrynoides spp.		
Nimbaphrynoides spp.		
Spinophrynoides spp.		
Dendrobatidae Poison frogs		
	Allobates femoralis	
	Cryptophyllobates azureiventris	
	Allobates zaparo	
	Dendrobates spp.	
	Epipedobates spp.	
	Phyllobates spp.	
Mantellidae Mantellas		
	Mantella spp.	
Microhylidae Red rain frog, tomato frog		
Dyscophus antongilii		
	Scaphiophryne gottlebei	
Rheobatrachidae Gastric-brooding frogs		
	Rheobatrachus spp.	
Ranidae Frogs		
	Euphlyctis hexadactylus	

(Continued)

	Hoplobatrachus tigerinus	
CAUDATA		
Ambystomatidae Axolotls		
	Ambystoma dumerilii	
	Ambystoma mexicanum	
Cryptobranchidae Giant salamanders		
Andrias **spp.**		
CLASS ELASMOBRANCHII (SHARKS)		
LAMNIFORMES		
Cetorhinidae Basking shark		
	Cetorhinus maximus	
Lamnidae Great white shark		
	Carcharodon carcharias	
ORECTOLOBIFORMES		
Rhincodontidae Whale shark		
	Rhincodon typus	
RAJIFORMES		
Pristidae Sawfishes		
Pristidae spp. (Except the species included in Appendix II)		
	Pristis microdon (For the exclusive purpose of allowing international trade in live animals to appropriate and acceptable aquaria for primarily conservation purposes)	
CLASS ACTINOPTERYGII (FISHES)		

(*Continued*)

(Continued)

ACIPENSERIFORMES Paddlefishes, sturgeons		
	ACIPENSERIFORMES spp. (Except the species included in Appendix I)	
Acipenseridae Sturgeons		
Acipenser brevirostrum		
Acipenser sturio		
ANGUILLIFORMES		
Anguillidae Freshwater eels		
	Anguilla anguilla	
CYPRINIFORMES		
Catostomidae Cui-ui		
Chasmistes cujus		
Cyprinidae Blind carps, plaeesok		
	Caecobarbus geertsi	
Probarbus jullieni		
OSTEOGLOSSIFORMES		
Osteoglossidae Arapaima, bonytongue		
	Arapaima gigas	
Scleropages formosus		
PERCIFORMES		
Labridae Wrasses		
	Cheilinus undulatus	
Sciaenidae Totoaba		
Totoaba macdonaldi		
SILURIFORMES		
Pangasiidae Pangasid catfish		

(Continued)

Pungasianodon gigas		
SYNGNATHIFORMES		
Syngnathidae Pipefishes, seahorses		
	Hippocampus **spp.**	
CLASS SARCOPTERYGII (LUNGFISHES)		
CERATODONTIFORMES		
Ceratodontidae Australian lungfish		
	Neoceratodus forsteri	
COELACANTHIFORMES		
Latimeriidae Coelacanths		
Latimeria **spp.**		
PHYLUM ECHINODERMATA CLASS HOLOTHUROIDEA (SEA CUCUMBERS)		
ASPIDOCHIROTIDA		
Stichopodidae Sea cucumbers		
		Isostichopus fuscus (Ecuador)
PHYLUM ARTHROPODA CLASS ARACHNIDA (SCORPIONS AND SPIDERS)		
ARANEAE		
Theraphosidae Red-kneed tarantulas, tarantulas		
	Aphonopelma albiceps	
	Aphonopelma pallidum	
	Brachypelma **spp.**	
SCORPIONES		

(*Continued*)

(Continued)

Scorpionidae Scorpions		
	Pandinus dictator	
	Pandinus gambiensis	
	Pandinus imperator	
CLASS INSECTA **(INSECTS)**		
COLEOPTERA		
Lucanidae Cape stag beetles		
		Colophon **spp.** (South Africa)
LEPIDOPTERA		
Papilionidae Birdwing butterflies, swallowtail butterflies		
	Atrophaneura jophon	
	Atrophaneura pandiyana	
	Bhutanitis **spp.**	
	Ornithoptera **spp.** (Except the species included in Appendix I)	
Ornithoptera alexandrae		
Papilio chikae		
Papilio homerus		
Papilio hospiton		
	Parnassius apollo	
	Teinopalpus **spp.**	
	Trogonoptera **spp.**	
	Troides **spp.**	
PHYLUM ANNELIDA **CLASS HIRUDINOIDEA** **(LEECHES)**		
ARHYNCHOBDELLIDA		

(Continued)

Hirudinidae Medicinal leech		
	Hirudo medicinalis	
PHYLUM MOLLUSCA **CLASS BIVALVIA** **(CLAMS AND MUSSELS)**		
MYTILOIDA		
Mytilidae Marine mussels		
	Lithophaga lithophaga	
UNIONOIDA		
Unionidae Freshwater mussels, pearly mussels		
Conradilla caelata		
	Cyprogenia aberti	
Dromus dromas		
Epioblasma curtisi		
Epioblasma florentina		
Epioblasma sampsonii		
Epioblasma sulcata perobliqua		
Epioblasma torulosa gubernaculum		
	Epioblasma torulosa rangiana	
Epioblasma torulosa torulosa		
Epioblasma turgidula		
Epioblasma walkeri		
Fusconaia cuneolus		
Fusconaia edgariana		
Lampsilis higginsii		

(*Continued*)

(Continued)

Lampsilis orbiculata orbiculata		
Lampsilis satur		
Lampsilis virescens		
Plethobasus cicatricosus		
Plethobasus cooperianus		
	Pleurobema clava	
Pleurobema plenum		
Potamilus capax		
Quadrula intermedia		
Quadrula sparsa		
Toxolasma cylindrella		
Unio nickliniana		
Unio tampicoensis tecomatensis		
Villosa trabalis		
VENEROIDA		
Tridacnidae Giant clams		
	Tridacnidae spp.	
CLASS GASTROPODA (SNAILS AND CONCHES)		
ARCHAEOGASTROPODA		
Haliotidae Abalones		
		Haliotis midae (South Africa)
MESOGASTROPODA		
Strombidae Queen conch		
	Strombus gigas	

(Continued)

STYLOMMATOPHORA		
Achatinellidae Agate snails, oahu tree snails		
***Achatinella* spp.**		
Camaenidae Green tree snail		
	Papustyla pulcherrima	
PHYLUM CNIDARIA CLASS ANTHOZOA (CORALS AND SEA ANEMONES)		
ANTIPATHARIA Black corals		
	ANTIPATHARIA spp.	
GORGONACEAE		
Coralliidae		
		Corallium elatius (China) *Corallium japonicum* (China) *Corallium konjoi* (China) **Corallium secundum** (China)
HELIOPORACEA		
Helioporidae Blue corals		
	Helioporidae spp. (Includes only the species *Heliopora coerulea*. Fossils are not subject to the provisions of the Convention)	
SCLERACTINIA Stony corals		
	SCLERACTINIA spp. (Fossils are not subject to the provisions of the Convention)	
STOLONIFERA		

(*Continued*)

(Continued)

Tubiporidae Organ-pipe corals		
	Tubiporidae spp. (Fossils are not subject to the provisions of the Convention)	
CLASS HYDROZOA **(SEA FERNS, FIRE CORALS AND STINGING MEDUSAE)**		
MILLEPORINA		
Milleporidae Fire corals		
	Milleporidae spp. (Fossils are not subject to the provisions of the Convention)	
STYLASTERINA		
Stylasteridae Lace corals		
	Stylasteridae spp. (Fossils are not subject to the provisions of the Convention)	
FLORA (PLANTS)		
AGAVACEAE Agaves		
Agave parviflora		
	Agave victoriae-reginae[1]	
	Nolina interrata	
AMARYLLIDACEAE Snowdrops, sternbergias		
	Galanthus **spp.**[1]	
	Sternbergia **spp.**[1]	
APOCYNACEAE Elephant trunks, hoodias		
	Hoodia **spp.**[9]	
	Pachypodium **spp.**[1] (Except the species included in Appendix I)	
Pachypodium ambongense		
Pachypodium baronii		

(Continued)

Pachypodium decaryi		
	Rauvolfia serpentina[#2]	
ARALIACEAE Ginseng		
	Panax ginseng[#3] Only the population of the Russian Federation; no other population is included in the Appendices)	
	Panax quinquefolius[#3]	
ARAUCARIACEAE Monkey-puzzle tree		
Araucaria araucana		
BERBERIDACEAE May-apple		
	Podophyllum hexandrum[#2]	
BROMELIACEAE Air plants, bromelias		
	Tillandsia harrisii[#1]	
	Tillandsia kammii[#1]	
	Tillandsia kautskyi[#1]	
	Tillandsia mauryana[#1]	
	Tillandsia sprengeliana[#1]	
	Tillandsia sucrei[#1]	
	Tillandsia xerographica[#1]	
CACTACEAE Cacti		
	CACTACEAE spp.[6, #4] (Except the species included in Appendix I and except *Pereskia* spp., *Pereskiopsis* spp. and *Quiabentia* spp.)	
Ariocarpus **spp.**		
Astrophytum asterias		
Aztekium ritteri		

(Continued)

(Continued)

Coryphantha werdermannii		
Discocactus spp.		
Echinocereus ferreirianus ssp. *lindsayi*		
Echinocereus schmollii		
Escobaria minima		
Escobaria sneedii		
Mammillaria pectinifera		
Mammillaria solisioides		
Melocactus conoideus		
Melocactus deinacanthus		
Melocactus glaucescens		
Melocactus paucispinus		
Obregonia denegrii		
Pachycereus militaris		
Pediocactus bradyi		
Pediocactus knowltonii		
Pediocactus paradinei		
Pediocactus peeblesianus		
Pediocactus sileri		
Pelecyphora spp.		
Sclerocactus brevihamatus ssp. *tobuschii*		
Sclerocactus erectocentrus		

(Continued)

Sclerocactus glaucus		
Sclerocactus mariposensis		
Sclerocactus mesae-verdae		
Sclerocactus nyensis		
Sclerocactus papyracanthus		
Sclerocactus pubispinus		
Sclerocactus wrightiae		
Strombocactus **spp.**		
Turbinicarpus **spp.**		
Uebelmannia **spp.**		
CARYOCARACEAE Ajo		
	Caryocar costaricense[1]	
COMPOSITAE (Asteraceae) Kuth		
Saussurea costus		
CRASSULACEAE Dudleyas		
	Dudleya stolonifera	
	Dudleya traskiae	
CUPRESSACEAE Alerce, cypresses		
Fitzroya cupressoides		
Pilgerodendron uviferum		
CYATHEACEAE Tree-ferns		
	Cyathea **spp.**[1]	
CYCADACEAE Cycads		

(*Continued*)

(Continued)

	CYCADACEAE spp.[#1] (Except the species included in Appendix I)	
Cycas beddomei		
DICKSONIACEAE Tree-ferns		
	Cibotium barometz[#1]	
	Dicksonia **spp.**[#1] (Only the populations of the Americas; no other population is included in the Appendices)	
DIDIEREACEAE Alluaudias, didiereas		
	DIDIEREACEAE spp.[#1]	
DIOSCOREACEAE Elephant's foot, kniss		
	Dioscorea deltoidea[#1]	
DROSERACEAE Venus' flytrap		
	Dionaea muscipula[#1]	
EUPHORBIACEAE Spurges		
	Euphorbia **spp.**[#1] (Succulent species only except the species included in Appendix I. Artificially propagated specimens of cultivars of *Euphorbia trigona*, artificially propagated specimens of crested, fan-shaped or colour mutants of *Euphorbia lactea*, when grafted on artificially propagated root stock of *Euphorbia neriifolia*, and artificially propagated specimens of cultivars of *Euphorbia* "Milii" when they are traded in shipments of 100 or more plants and readily recognizable as artificially propagated specimens, are not subject to the provisions of the Convention)	

(Continued)

Euphorbia ambovombensis		
Euphorbia capsaintemariensis		
Euphorbia cremersii (Includes the *forma viridifolia* and the var. *rakotozafyi*)		
Euphorbia cylindrifolia (Includes the ssp. *tuberifera*)		
Euphorbia decaryi (Includes the vars. *ampanihyensis*, *robinsonii* and *spirosticha*)		
Euphorbia francoisii		
Euphorbia moratii (Includes the vars. *antsingiensis*, *bemarahensis* and *multiflora*)		
Euphorbia parvicyathophora		
Euphorbia quartziticola		
Euphorbia tulearensis		
FOUQUIERIACEAE Ocotillos		
	Fouquieria columnaris[1]	
Fouquieria fasciculata		
Fouquieria purpusii		
GNETACEAE Gnetums		
		Gnetum montanum[1] (Nepal)
JUGLANDACEAE Gavilan		

(Continued)

(Continued)

	Oreomunnea pterocarpa[#1]	
LEGUMINOSAE (Fabaceae) Afrormosia, cristobal, rosewood, sandalwood		
	Caesalpinia echinata[#10]	
Dalbergia nigra		
		Dalbergia retusa[#5] (population of Guatemala [Guatemala]) *Dalbergia stevensonii*[#5] (population of Guatemala [Guatemala]) *Dipteryx panamensis* (Costa Rica, Nicaragua)
	Pericopsis elata[#5]	
	Platymiscium pleiostachyum[#1]	
	Pterocarpus santalinus[#7]	
LILIACEAE Aloes		
	Aloe **spp.**[#1] (Except the species included in Appendix I. Also excludes *Aloe vera*, also referenced as *Aloe barbadensis* which is not included in the Appendices)	
Aloe albida		
Aloe albiflora		
Aloe alfredii		
Aloe bakeri		
Aloe bellatula		
Aloe calcairophila		
Aloe compressa (Includes the vars. *paucituberculata,*		

(Continued)

rugosquamosa and *schistophila*)		
Aloe delphinensis		
Aloe descoingsii		
Aloe fragilis		
Aloe haworthioides (Includes the var. *aurantiaca*)		
Aloe helenae		
Aloe laeta (Includes the var. *maniaensis*)		
Aloe parallelifolia		
Aloe parvula		
Aloe pillansii		
Aloe polyphylla		
Aloe rauhii		
Aloe suzannae		
Aloe versicolor		
Aloe vossii		
MAGNOLIACEAE Magnolia		
		Magnolia liliifera var. ***obovata***[1] (Nepal)
MELIACEAE Mahoganies, Spanish cedar		
		Cedrela odorata[5] (Population of Colombia [Colombia] Population of Guatemala [Guatemala] Population of Peru [Peru])
	Swietenia humilis[1]	

(*Continued*)

(Continued)

	Swietenia macrophylla[#6] (Populations of the Neotropics)	
	Swietenia mahagoni[#5]	
NEPENTHACEAE Pitcher-plants (Old World)		
	Nepenthes spp.[#1] (Except the species included in Appendix I)	
Nepenthes khasiana		
Nepenthes rajah		
ORCHIDACEAE Orchids		
	ORCHIDACEAE spp. [7, #1] (Except the species included in Appendix I)	
(For all of the following Appendix-I species, seedling or tissue cultures obtained *in vitro*, in solid or liquid media, transported in sterile containers are not subject to the provisions of the Convention)		
Aerangis ellisii		
Dendrobium cruentum		
Laelia jongheana		
Laelia lobata		
Paphiopedilum **spp.**		
Peristeria elata		
Phragmipedium **spp.**		
Renanthera imschootiana		
OROBANCHACEAE Broomrape		

(Continued)

	Cistanche deserticola[1]	
PALMAE (Arecaceae) Palms		
	Beccariophoenix madagascariensis[1]	
Chrysalidocarpus decipiens		
	Lemurophoenix halleuxii	
	Marojejya darianii	
	Neodypsis decaryi[1]	
	Ravenea louvelii	
	Ravenea rivularis	
	Satranala decussilvae	
	Voanioala gerardii	
PAPAVERACEAE Poppy		
		Meconopsis regia[1] (Nepal)
PINACEAE Guatemala fir		
Abies guatemalensis		
PODOCARPACEAE Podocarps		
		Podocarpus neriifolius[1] (Nepal)
Podocarpus parlatorei		
PORTULACACEAE Lewisias, portulacas, purslanes		
	Anacampseros spp.[1]	
	Avonia spp.[1]	
	Lewisia serrata[1]	
PRIMULACEAE Cyclamens		

(*Continued*)

(Continued)

	Cyclamen **spp.**[8, #1]	
PROTEACEAE Proteas		
	Orothamnus zeyheri[#1]	
	Protea odorata[#1]	
RANUNCULACEAE Golden seals, yellow adonis, yellow root		
	Adonis vernalis[#2]	
	Hydrastis canadensis[#8]	
ROSACEAE African cherry, stinkwood		
	Prunus africana[#1]	
RUBIACEAE Ayugue		
Balmea stormiae		
SARRACENIACEAE Pitcher-plants (New World)		
	Sarracenia **spp.**[#1] (Except the species included in Appendix I)	
Sarracenia oreophila		
Sarracenia rubra ssp. *alabamensis*		
Sarracenia rubra ssp. *jonesii*		
SCROPHULARIACEAE Kutki		
	Picrorhiza kurrooa[#2] (Excludes *Picrorhiza scrophulariiflora*)	
STANGERIACEAE Stangerias		
	Bowenia **spp.**[#1]	
Stangeria eriopus		
TAXACEAE Himalayan yew		
	Taxus chinensis and infraspecific taxa of this species[#2]	

(Continued)

	Taxus cuspidata and infraspecific taxa of this species[9, #2]	
	Taxus fuana and infraspecific taxa of this species[#2]	
	Taxus sumatrana and infraspecific taxa of this species[#2]	
	Taxus wallichiana[#2]	
THYMELAEACEAE (Aquilariaceae) Agarwood, ramin		
	Aquilaria spp.[#1]	
	Gonystylus spp.[#1]	
	Gyrinops spp.[#1]	
TROCHODENDRACEAE (Tetracentraceae) Tetracentron		
		Tetracentron sinense[#1] (Nepal)
VALERIANACEAE Himalayan spikenard		
	Nardostachys grandiflora[#2]	
WELWITSCHIACEAE Welwitschia		
	Welwitschia mirabilis[#1]	
ZAMIACEAE Cycads		
	ZAMIACEAE spp.[#1] (Except the species included in Appendix I)	
Ceratozamia spp.		
Chigua spp.		
Encephalartos spp.		
Microcycas calocoma		
ZINGIBERACEAE Ginger lily		
	Hedychium philippinense[#1]	

(*Continued*)

(Continued)

ZYGOPHYLLACEAE Lignum-vitae		
		Bulnesia sarmientoi[#11] (Argentina)
	Guaiacum spp.[#2]	

[1] *Population of Argentina (listed in Appendix II):*

For the exclusive purpose of allowing international trade in wool sheared from live vicuñas, in cloth, and in derived manufactured products and other handicraft artefacts. The reverse side of the cloth must bear the logotype adopted by the range States of the species, which are signatories to the Convenio para la Conservación y Manejo de la Vicuña, and the selvages the words "VICUÑA-ARGENTINA." Other products must bear a label including the logotype and the designation "VICUÑA-ARGENTINA-ARTESANÍA."

All other specimens shall be deemed to be specimens of species included in Appendix I and the trade in them shall be regulated accordingly.

[2] *Population of Bolivia (listed in Appendix II):*

For the exclusive purpose of allowing international trade in wool sheared from live vicuñas, and in cloth and items made thereof, including luxury handicrafts and knitted articles.

The reverse side of the cloth must bear the logotype adopted by the range States of the species, which are signatories to the Convenio para la Conservación y Manejo de la Vicuña, and the selvages the words "VICUÑA-BOLIVIA." Other products must bear a label including the logotype and the designation "VICUÑA-BOLIVIA-ARTESANÍA."

All other specimens shall be deemed to be specimens of species included in Appendix I and the trade in them shall be regulated accordingly.

[3] *Population of Chile (listed in Appendix II):*

For the exclusive purpose of allowing international trade in wool sheared from live vicuñas, and in cloth and items made thereof, including luxury handicrafts and knitted articles. The reverse side of the cloth must bear the logotype adopted by the range States of the species, which are signatories to the Convenio para la Conservación y Manejo de la Vicuña, and the selvages the words "VICUÑA-CHILE." Other products must bear a label including the logotype and the designation "VICUÑA-CHILE-ARTESANÍA."

All other specimens shall be deemed to be specimens of species included in Appendix I and the trade in them shall be regulated accordingly.

[4] *Population of Peru (listed in Appendix II):*

For the exclusive purpose of allowing international trade in wool sheared from live vicuñas and in the stock extant at the time of the ninth meeting of the Conference of the Parties (November 1994) of 3249 kg of wool, and in cloth and items made thereof, including luxury handicrafts and knitted articles. The reverse side of the cloth must bear the logotype adopted by the range States of the species, which are signatories to the Convenio para la Conservación y Manejo de la Vicuña, and the selvages the words "VICUÑA-PERÚ." Other products must bear a label including the logotype and the designation "VICUÑA-PERÚ-ARTESANÍA."

All other specimens shall be deemed to be specimens of species included in Appendix I and the trade in them shall be regulated accordingly.

[5] *Populations of Botswana, Namibia, South Africa and Zimbabwe (listed in Appendix II):*

For the exclusive purpose of allowing:

a) trade in hunting trophies for non-commercial purposes;

b) trade in live animals to appropriate and acceptable destinations, as defined in Resolution Conf. 11.20, for Botswana and Zimbabwe and for in situ conservation programmes for Namibia and South Africa;

c) trade in hides;

d) trade in hair;

e) trade in leather goods for commercial or non-commercial purposes for Botswana, Namibia and South Africa and for non-commercial purposes for Zimbabwe;

f) trade in individually marked and certified ekipas incorporated in finished jewellery for non-commercial purposes for Namibia and ivory carvings for non-commercial purposes for Zimbabwe;

g) trade in registered raw ivory (for Botswana, Namibia, South Africa and Zimbabwe, whole tusks and pieces) subject to the following:

i) only registered government-owned stocks, originating in the State (excluding seized ivory and ivory of unknown origin);

ii) only to trading partners that have been verified by the Secretariat, in consultation with the Standing Committee, to have sufficient national legislation and domestic trade controls to ensure that the imported ivory will not be re-exported and will be managed in accordance with all requirements of Resolution Conf. 10.10 (Rev. CoP14) concerning domestic manufacturing and trade;

iii) not before the Secretariat has verified the prospective importing countries and the registered government-owned stocks;

iv) raw ivory pursuant to the conditional sale of registered government-owned ivory stocks agreed at CoP12, which are 20,000 kg (Botswana), 10,000 kg (Namibia) and 30,000 kg (South Africa);

v) in addition to the quantities agreed at CoP12, government-owned ivory from Botswana, Namibia, South Africa and Zimbabwe registered by 31 January 2007 and verified by the Secretariat may be traded and despatched, with the ivory in paragraph g) iv) above, in a single sale per destination under strict supervision of the Secretariat;

vi) the proceeds of the trade are used exclusively for elephant conservation and community conservation and development programmes within or adjacent to the elephant range; and vii) the additional quantities specified in paragraph g) v) above shall be traded only after the Standing Committee has agreed that the above conditions have been met; and

h) no further proposals to allow trade in elephant ivory from populations already in Appendix II shall be submitted to the Conference of the Parties for the period from CoP14 and ending nine years from the date of the single sale of ivory that is to take place in accordance with provisions in paragraphs g) i), g) ii), g) iii), g) vi) and g) vii). In addition such further proposals shall be dealt with in accordance with Decisions 14.77 and 14.78.

On a proposal from the Secretariat, the Standing Committee can decide to cause this trade to cease partially or completely in the event of non-compliance by exporting or importing countries, or in the case of proven detrimental impacts of the trade on other elephant populations.

All other specimens shall be deemed to be specimens of species included in Appendix I and the trade in them shall be regulated accordingly.

[6]Artificially propagated specimens of the following hybrids and/or cultivars are not subject to the provisions of the Convention:

– *Hatiora x graeseri*

– *Schlumbergera x buckleyi*

– *Schlumbergera russelliana x Schlumbergera truncata*

– *Schlumbergera orssichiana x Schlumbergera truncata*

– *Schlumbergera opuntioides x Schlumbergera truncata*

– *Schlumbergera truncata (cultivars)*

– Cactaceae spp. colour mutants lacking chlorophyll, grafted on the following grafting stocks: *Harrisia* "Jusbertii," *Hylocereus trigonus* or *Hylocereus undatus*

– *Opuntia microdasys* (cultivars).

[7]Artificially propagated hybrids of the following genera are not subject to the provisions of the Convention, if conditions, as indicated under a) and b), are met: *Cymbidium, Dendrobium, Phalaenopsis* and *Vanda*:

   a) Specimens are readily recognizable as artificially propagated and do not show any signs of having been collected in the wild such as mechanical damage or strong dehydration resulting from collection, irregular growth and heterogeneous size and shape within a taxon and shipment, algae or other epiphyllous organisms adhering to leaves, or damage by insects or other pests; and
   b)

      i) when shipped in non-flowering state, the specimens must be traded in shipments consisting of individual containers (such as cartons, boxes, crates or individual shelves of CC-containers) each containing 20 or more plants of the same hybrid; the plants within each container must exhibit a high degree of uniformity and healthiness; and the shipment must be accompanied by documentation, such as an invoice, which clearly states the number of plants of each hybrid; or
      ii) when shipped in flowering state, with at least one fully open flower per specimen, no minimum number of specimens per shipment is required but specimens must be professionally processed for commercial retail sale, e.g., labelled with printed labels or packaged with printed packages indicating the name of the hybrid and the country of final processing. This should be clearly visible and allow easy verification.

   Plants not clearly qualifying for the exemption must be accompanied by appropriate CITES documents.

[8]Artificially propagated specimens of *cultivars of Cyclamen persicum* are not subject to the provisions of the Convention. However, the exemption does not apply to such specimens traded as dormant tubers.

[9]Artificially propagated hybrids and cultivars of *Taxus cuspidata*, live, in pots or other small containers, each consignment being accompanied by a label or document stating the name of the taxon or taxa and the text "artificially propagated," are not subject to the provisions of the Convention.

[#1]All parts and derivatives, except:

   a) seeds, spores and pollen (including pollinia);
   b) seedling or tissue cultures obtained *in vitro*, in solid or liquid media, transported in sterile containers;
   c) cut flowers of artificially propagated plants; and
   d) fruits and parts and derivatives thereof of artificially propagated plants of the genus *Vanilla*.

[#2]All parts and derivatives except:

   a) seeds and pollen; and
   b) finished products packaged and ready for retail trade.

[#3]Whole and sliced roots and parts of roots.

[#4]All parts and derivatives, except:

   a) seeds, except those from Mexican cacti originating in Mexico, and pollen;
   b) seedling or tissue cultures obtained in vitro, in solid or liquid media, transported in sterile containers;
   c) cut flowers of artificially propagated plants;
   d) fruits and parts and derivatives thereof of naturalized or artificially propagated plants; and
   e) separate stem joints (pads) and parts and derivatives thereof of naturalized or artificially propagated plants of the genus *Opuntia* subgenus *Opuntia*.

[5]Logs, sawn wood and veneer sheets.

[6]Logs, sawn wood, veneer sheets and plywood.

[7]Logs, wood-chips, powder and extracts.

[8]Underground parts (i.e., roots, rhizomes): whole, parts and powdered.

[9]All parts and derivatives except those bearing a label

"Produced from Hoodia spp. material obtained through controlled harvesting and production in collaboration with the CITES Management Authorities of Botswana/Namibia/South Africa under agreement no. BW/NA/ZA xxxxxx."

[10]Logs, sawn wood, veneer sheets, including unfinished wood articles used for the fabrication of bows for stringed musical instruments.

[11]Logs, sawn wood, veneer sheets, plywood, powder and extracts.

Appendix C
The Lacey Act

United States Code Annotated. Title 16. Conservation. Chapter 53. Control of Illegally Taken Fish and Wildlife.

Citation: 16 USC 3371–3378

Citation: 95 Stat. 1073

Summary:

The Lacey Act provides that it is unlawful for any person to import, export, transport, sell, receive, acquire, or purchase any fish or wildlife or plant taken, possessed, transported, or sold in violation of any law, treaty, or regulation of the United States or in violation of any Indian tribal law whether in interstate or foreign commerce. Violation of this federal act can result in civil penalties up to $10,000 per each violation or maximum criminal sanctions of $20,000 in fines and/or up to five years imprisonment. All plants or animals taken in violation of the Act are subject to forfeiture as well as all vessels, vehicles, aircraft, and other equipment used to aid in the importing, exporting, transporting, selling, receiving, acquiring, or purchasing of fish or wildlife or plants in a criminal violation of this chapter for which a felony conviction is obtained where the owner should have known of the illegal transgression.

§ 3371. DEFINITIONS

For the purposes of this chapter:

(a) The term "fish or wildlife" means any wild animal, whether alive or dead, including without limitation any wild mammal, bird, reptile, amphibian, fish, mollusk, crustacean, arthropod, coelenterate, or other invertebrate, whether or not bred, hatched, or born in captivity, and includes any part, product, egg, or offspring thereof.

(b) The term "import" means to land on, bring into, or introduce into, any place subject to the jurisdiction of the United States, whether or not such landing, bringing, or introduction constitutes an importation within the meaning of the customs laws of the United States.

(c) The term "Indian tribal law" means any regulation of, or other rule of conduct enforceable by, any Indian tribe, band, or group but only to the extent that the regulation or rule applies within Indian country as defined in section 1151 of Title 18.

(d) The terms "law," "treaty," "regulation," and "Indian tribal law" mean laws, treaties, regulations or Indian tribal laws which regulate the taking, possession, importation, exportation, transportation, or sale of fish or wildlife or plants.

(e) The term "person" includes any individual, partnership, association, corporation, trust, or any officer, employee, agent, department, or instrumentality of the Federal Government or of any State or political subdivision thereof, or any other entity subject to the jurisdiction of the United States.

(f) Plant

   (1) In general

      The terms "plant" and "plants" mean any wild member of the plant kingdom, including roots, seeds, parts, or products thereof, and including trees from either natural or planted forest stands.

   (2) Exclusions

      The terms "plant" and "plants" exclude—

   (A) common cultivars, except trees, and common food crops (including roots, seeds, parts, or products thereof);

   (B) a scientific specimen of plant genetic material (including roots, seeds, germplasm, parts, or products thereof) that is to be used only for laboratory or field research; and

   (C) any plant that is to remain planted or to be planted or replanted.

   (3) Exceptions to application of exclusions

      The exclusions made by subparagraphs (B) and (C) of paragraph (2) do not apply if the plant is listed—

   (A) in an appendix to the Convention on International Trade in Endangered Species of Wild Fauna and Flora (27 UST 1087; TIAS 8249);

   (B) as an endangered or threatened species under the Endangered Species Act of 1973 (16 U.S.C. 1531 et seq.); or

   (C) pursuant to any State law that provides for the conservation of species that are indigenous to the State and are threatened with extinction.

(g) Prohibited wildlife species—The term "prohibited wildlife species" means any live species of lion, tiger, leopard, cheetah, jaguar, or cougar or any hybrid of such a species.

(h) The term "Secretary" means, except as otherwise provided in this chapter, the Secretary of the Interior or the Secretary of Commerce, as program responsibilities are vested pursuant to the provisions of Reorganization Plan

Numbered 4 of 1970(84 Stat. 2090); except that with respect to the provisions of this chapter which pertain to the importation or exportation of plants, the term also means the Secretary of Agriculture.

(i) The term "State" means any of the several States, the District of Columbia, the Commonwealth of Puerto Rico, the Virgin Islands, Guam, Northern Mariana Islands, American Samoa, and any other territory, commonwealth, or possession of the United States.

(j) Taken and taking

   (1) Taken
   The term "taken" means captured, killed, or collected and, with respect to a plant, also means harvested, cut, logged, or removed.

   (2) Taking
   The term "taking" means the act by which fish, wildlife, or plants are taken.

(k) The term "transport" means to move, convey, carry, or ship by any means, or to deliver or receive for the purpose of movement, conveyance, carriage, or shipment.

CREDIT(S)

(Pub.L. 97-79, § 2, Nov. 16, 1981, 95 Stat. 1073; Pub.L. 108-191, § 2, Dec. 19, 2003, 117 Stat. 2871; Pub.L. 110-234, Title VIII, § 8204(a) (1), (2), May 22, 2008, 122 Stat. 1291; Pub.L. 110-246, Title VIII,§ 8204(a), June 18, 2008, 122 Stat. 2052.)

§ 3372. PROHIBITED ACTS

(a) Offenses other than marking offenses
   It is unlawful for any person—

   (1) to import, export, transport, sell, receive, acquire, orpurchase any fish or wildlife or plant taken, possessed, transported, or sold in violation of any law, treaty, or regulation of the United States or in violation of any Indian tribal law;

   (2) to import, export, transport, sell, receive, acquire, orpurchase in interstate or foreign commerce—

   (A) any fish or wildlife taken, possessed, transported, or sold in violation of any law or regulation of any State or in violation of any foreign law;

   (B) any plant—

   (i) taken, possessed, transported, or sold in violation of any law or regulation of any State, or any foreign law, that protects plants or that regulates—

   (I) the theft of plants;

   (II) the taking of plants from a park, forest reserve, or otherofficially protected area;

   (III) the taking of plants from an officially designated area; or

   (IV) the taking of plants without, or contrary to, required authorization;

   (ii) taken, possessed, transported, or sold without the paymentof appropriate royalties, taxes, or stumpage fees requiredfor the plant by any law or regulation of any State or anyforeign law; or

(iii) taken, possessed, transported, or sold in violation of any limitation under any law or regulation of any State, or under any foreign law, governing the export or transshipment of plants; or

(C) any prohibited wildlife species (subject to subsection (e) of this section);

(3) within the special maritime and territorial jurisdiction of the United States (as defined in section 7 of Title 18)—

(A) to possess any fish or wildlife taken, possessed, transported, or sold in violation of any law or regulation of any State or in violation of any foreign law or Indian tribal law, or

(B) to possess any plant—

(i) taken, possessed, transported, or sold in violation of any law or regulation of any State, or any foreign law, that protects plants or that regulates—

(I) the theft of plants;

(II) the taking of plants from a park, forest reserve, or otherofficially protected area;

(III) the taking of plants from an officially designated area; or

(IV) the taking of plants without, or contrary to, required authorization;

(ii) taken, possessed, transported, or sold without the payment of appropriate royalties, taxes, or stumpage fees required for the plant by any law or regulation of any State or any foreign law; or

(iii) taken, possessed, transported, or sold in violation of any limitation under any law or regulation of any State, or under any foreign law, governing the export or transshipment of plants; or

(4) to attempt to commit any act described in paragraphs(1) through (3).

(b) Marking offenses

It is unlawful for any person to import, export, or transport in interstate commerce any container or package containing any fish or wildlife unless the container or package has previously been plainly marked, labeled, or tagged in accordance with the regulations issued pursuant to paragraph (2) of section 3376(a) of this title.

(c) Sale and purchase of guiding and outfitting services and invalid licenses and permits

(1) Sale

It is deemed to be a sale of fish or wildlife in violation of this chapter for a person for money or other consideration to offer or provide—

(A) guiding, outfitting, or other services; or

(B) a hunting or fishing license or permit;
for the illegal taking, acquiring, receiving, transporting, or possessing of fish or wildlife.

(2) Purchase

It is deemed to be a purchase of fish or wildlife in violation of this chapter for a person to obtain for money or other consideration—

(A) guiding, outfitting, or other services; or

(B) a hunting or fishing license or permit;
for the illegal taking, acquiring, receiving, transporting, or possessing of fish or wildlife.

(d) False labeling offenses

It is unlawful for any person to make or submit any false record, account, or label for, or any false identification of, any fish, wildlife, or plant which has been, or is intended to be—

(1) imported, exported, transported, sold, purchased, or received from any foreign country; or

(2) transported in interstate or foreign commerce.

(e) Nonapplicability of prohibited wildlife species offense

(1) In general

Subsection (a)(2)(C) of this section does not apply to importation, exportation, transportation, sale, receipt, acquisition, or purchase of an animal of a prohibited wildlife species, by a person that, under regulations prescribed under paragraph (3), is described in paragraph (2) with respect to that species.

(2) Persons described

A person is described in this paragraph, if the person—

(A) is licensed or registered, and inspected, by the Animal and Plant Health Inspection Service or any other Federal agency with respect to that species;

(B) is a State college, university, or agency, State-licensedwildlife rehabilitator, or State-licensed veterinarian;

(C) is an accredited wildlife sanctuary that cares for prohibited wildlife species and—

(i) is a corporation that is exempt from taxation under section 501(a) of Title 26 and described in sections 501(c)(3) and 170(b)(1)(A)(vi) of Title 26;

(ii) does not commercially trade in animals listed in section 3371(g) of this title, including offspring, parts, and byproducts of such animals;

(iii) does not propagate animals listed in section 3371(g) of this title; and

(iv) does not allow direct contact between the public andanimals; or

(D) has custody of the animal solely for the purpose ofexpeditiously transporting the animal to a person described in this paragraph with respect to the species.

(3) Regulations

Not later than 180 days after December 19, 2003, theSecretary, in cooperation with the Director of the Animal and Plant Health Inspection Service, shall promulgateregulations describing the persons described in paragraph (2).

(4) State authority

Nothing in this subsection preempts or supersedes the authority of a State to regulate wildlife species within that State.

(5) Authorization of appropriations

There is authorized to be appropriated to carry out subsection (a)(2)(C) of this section $3,000,000 for each of fiscal years 2004 through 2008.

(f) Plant declarations

(1) Import declaration

Effective 180 days from the date of enactment of this subsection, and except as provided in paragraph (3), it shall be unlawful for any person to import any plant unless the person files upon importation a declaration that contains—

(A) the scientific name of any plant (including the genus andspecies of the plant) contained in the importation;

(B) a description of—

(i) the value of the importation; and

(ii) the quantity, including the unit of measure, of theplant; and

(C) the name of the country from which the plant was taken.

(2) Declaration relating to plant products
Until the date on which the Secretary promulgates a regulation under paragraph (6), a declaration relating to a plant product shall—

(A) in the case in which the species of plant used to produce the plant product that is the subject of the importation varies, and the species used to produce the plant product is unknown, contain the name of each species of plant that may have been used to produce the plant product;

(B) in the case in which the species of plant used to producethe plant product that is the subject of the importation is commonly taken from more than one country, and the country from which the plant was taken and used toproduce the plant product is unknown, contain the name of each country from which the plant may have been taken; and

(C) in the case in which a paper or paperboard plant product includes recycled plant product, contain the average percent recycled content without regard for the species or country of origin of the recycled plant product, in addition to theinformation for the non-recycled plant content otherwise required by this subsection.

(3) Exclusions
Paragraphs (1) and (2) shall not apply to plants usedexclusively as packaging material to support, protect, or carry another item, unless the packaging material itself is the item being imported.

(4) Review
Not later than two years after the date of enactment of this subsection, the Secretary shall review the implementation of each requirement imposed by paragraphs (1) and (2) and the effect of the exclusion provided by paragraph (3). Inconducting the review, the Secretary shall provide public notice and an opportunity for comment.

(5) Report
Not later than 180 days after the date on which the Secretary completes the review under paragraph (4), the Secretary shall submit to the appropriate committees of Congress a report containing—

(A) an evaluation of—

(i) the effectiveness of each type of information requiredunder paragraphs (1) and (2) in assisting enforcement of this section; and

(ii) the potential to harmonize each requirement imposedby paragraphs (1) and (2) with other applicable importregulations in existence as of the date of the report;

(B) recommendations for such legislation as the Secretarydetermines to be appropriate to assist in the identification of plants that are imported into the United States in violation of this section; and

(C) an analysis of the effect of subsection (a) and this subsection on—

(i) the cost of legal plant imports; and

(ii) the extent and methodology of illegal logging practices and trafficking.

(6) Promulgation of regulations
Not later than 180 days after the date on which the Secretary completes the review under paragraph (4), the Secretary may promulgate regulations—

(A) to limit the applicability of any requirement imposed by paragraph (2) to specific plant products;

(B) to make any other necessary modification to any requirement imposed by paragraph (2), as determined by the Secretary based on the review; and

(C) to limit the scope of the exclusion provided by paragraph(3), if the limitations in scope are warranted as a result ofthe review.

CREDIT(S)

(Pub.L. 97-79, § 3, Nov. 16, 1981, 95 Stat. 1074; Pub.L. 100-653, Title I, § 101, Nov. 14, 1988, 102 Stat. 3825; Pub.L. 108-191,§ 3(a), Dec. 19, 2003, 117 Stat. 2871; Pub.L. 110-234, Title VIII, § 8204(b), May 22, 2008, 122 Stat. 1292; Pub.L. 110-246, Title VIII, § 8204(b), June 18, 2008, 122 Stat. 2053.)

§ 3373. PENALTIES AND SANCTIONS

(a) Civil penalties

(1) Any person who engages in conduct prohibited by any provision of this chapter (other than subsections (b), (d) and (f) of section 3372 of this title) and in the exercise of due care should know that the fish or wildlife or plants were taken, possessed, transported, or sold in violation of, or in a manner unlawful under, any underlying law, treaty, or regulation, and any person who knowingly violates subsection (d) or (f) of section 3372 of this title, may be assessed a civil penalty by the Secretary of not more than $10,000 for each such violation: *Provided*, That when the violation involves fish or wildlife or plants with a market value of less than $350, and involves only the transportation, acquisition, or receipt of fish or wildlife or plants taken orpossessed in violation of any law, treaty, or regulation of the United States, any Indian tribal law, any foreign law, or any law or regulation of any State, the penalty assessed shall not exceed the maximum provided for violation of said law, treaty, or regulation, or $10,000, whichever is less.

(2) Any person who violates subsection (b) or (f) of section 3372 of this title, except as provided in paragraph (1), may be assessed a civil penalty by the Secretary of not more than $250.

(3) For purposes of paragraphs (1) and (2), any reference to a provision of this chapter or to a section of this chapter shall be treated as including any regulation issued to carry out any such provision or section.

(4) No civil penalty may be assessed under this subsection unless the person accused of the violation is given notice andopportunity for a hearing with

respect to the violation. Each violation shall be a separate offense and the offense shall be deemed to have been committed not only in the district where the violation first occurred, but also in any district in which a person may have taken or been in possession of the said fish or wildlife or plants.

(5) Any civil penalty assessed under this subsection may be remitted or mitigated by the Secretary.

(6) In determining the amount of any penalty assessed pursuant to paragraphs (1) and (2), the Secretary shall take into account the nature, circumstances, extent, and gravity of the prohibited act committed, and with respect to the violator, the degree of culpability, ability to pay, and such other matters as justice may require.

(b) Hearings

Hearings held during proceedings for the assessment of civil penalties shall be conducted in accordance with section 554 of Title 5. The administrative law judge may issue subpenas for the attendance and testimony of witnesses and the production of relevant papers, books, or documents, and may administer oaths. Witnesses summoned shall be paid the same fees and mileage that are paid to witnesses in the courts of the United States. In case of contumacy or refusal to obey a subpena issued pursuant to this paragraph and served upon any person, the district court of the United States for any district in which such person is found, resides, or transacts business, upon application by the United States and after notice to such person, shall have jurisdiction to issue an order requiring such person to appear and give testimony before the administrative law judge or to appear and produce documents before the administrative law judge, or both, and any failure to obey such order of the court may be punished by such court as a contempt thereof.

(c) Review of civil penalty

Any person against whom a civil penalty is assessed under this section may obtain review thereof in the appropriate District Court of the United States by filing a complaint in such court within 30 days after the date of such order and by simultaneously serving a copy of the complaint by certified mail on the Secretary, the Attorney General, and the appropriate United States attorney. The Secretary shall promptly file in such court a certified copy of the record upon which such violation was found or such penalty imposed, as provided in section 2112 of Title 28. If any person fails to pay an assessment of a civil penalty after it has become a final and unappealable order or after the appropriate court has entered final judgment in favor of the Secretary, the Secretary may request the Attorney General of the United States to institute a civil action in an appropriate district court of the United States to collect the penalty, and such court shall have jurisdiction to hear and decide any such action. In hearing such action, the court shall have authority to review the violation and the assessment of the civil penalty de novo.

(d) Criminal penalties

(1) Any person who—

(A) knowingly imports or exports any fish or wildlife or plantsin violation of any provision of this chapter (other thansubsections (b), (d) and (f) of section 3372 of this title), or

(B) violates any provision of this chapter (other than subsections (b), (d) and (f) of section 3372 of this title) by knowinglyengaging in conduct that involves the sale or purchase of, the offer of sale or purchase of, or the intent to sell or purchase, fish or wildlife or plants with a market value in excess of $350, knowing that the fish or wildlife or plants were taken, possessed, transported, or sold in violation of, or in a manner unlawful under, any underlying law, treaty or regulation, shall be fined not more than $20,000, or imprisoned for not more than five years, or both. Each violation shall be a separate offense and the offense shall be deemed to have been committed not only in the district where the violation first occurred, but also in any district in which the defendant may have taken or been in possession of the said fish or wildlife or plants.

(2) Any person who knowingly engages in conduct prohibited by any provision of this chapter (other than subsections (b), (d), and (f) of section 3372 of this title) and in the exercise of due care should know that the fish or wildlife or plants were taken, possessed, transported, or sold in violation of, or in a manner unlawful under, any underlying law, treaty or regulation shall be fined not more than $10,000, or imprisoned for not more than one year, or both. Each violation shall be a separate offense and the offense shall be deemed to have been committed not only in the district where the violation first occurred, but also in any district in which the defendant may have taken or been in possession of the said fish or wildlife or plants.

(3) Any person who knowingly violates subsection (d) or (f) of section 3372 of this title—

(A) shall be fined under Title 18, or imprisoned for not more than 5 years, or both, if the offense involves—

(i) the importation or exportation of fish or wildlife or plants; or

(ii) the sale or purchase, offer of sale or purchase, or commission of an act with intent to sell or purchase fish or wildlife or plants with a market value greater than $350; and

(B) shall be fined under Title 18, or imprisoned for not more than 1 year, or both, if the offense does not involve conduct described in subparagraph (A).

(e) Permit sanctions

The Secretary may also suspend, modify, or cancel any Federal hunting or fishing license, permit, or stamp, or any license or permit authorizing a person to import or export fish or wildlife or plants (other than a permit or license issued pursuant to the Magnuson-Stevens Fishery Conservation and Management Act [16 U.S.C.A. § 1801 et seq.]), or to operate a quarantine station or rescue center for imported wildlife or plants, issued to any person who is convicted of a criminal violation of any provision of this chapter or any regulation issued hereunder. The Secretary shall not be liable for the payments of any compensation, reimbursement, or damages in connection with the modification, suspension, or revocation of any licenses, permits, stamps, or other agreements pursuant to this section.

CREDIT(S)

(Pub.L. 97-79, § 4, Nov. 16, 1981, 95 Stat. 1074; Pub.L. 100-653, Title I, § 102, 103, Nov. 14, 1988, 102 Stat. 3825, 3826; Pub.L. 104-208, Div. A, Title I, § 101(a) [Title II, § 211(b)], Sept. 30, 1996, 110 Stat. 3009-41; Pub.L. 110-234, Title VIII, § 8204(c), (f), May 22,

2008, 122 Stat. 1294; Pub.L. 110-246, Title VIII, § 8204(c), (f), June 18, 2008, 122 Stat. 2055, 2056.)

§ 3374. FORFEITURE

(a) In general

(1) All fish or wildlife or plants imported, exported, transported, sold, received, acquired, or purchased contrary to theprovisions of section 3372 of this title (other than section 3372(b) of this title), or any regulation issued pursuant thereto, shall be subject to forfeiture to the United Statesnotwithstanding any culpability requirements for civil penalty assessment or criminal prosecution included in section 3373 of this title.

(2) All vessels, vehicles, aircraft, and other equipment used to aid in the importing, exporting, transporting, selling, receiving, acquiring, or purchasing of fish or wildlife or plants in acriminal violation of this chapter for which a felony conviction is obtained shall be subject to forfeiture to the United States if (A) the owner of such vessel, vehicle, aircraft, or equipment was at the time of the alleged illegal act a consenting party or privy thereto or in the exercise of due care should have known that such vessel, vehicle, aircraft, or equipment would be used in a criminal violation of this chapter, and (B) the violation involved the sale or purchase of, the offer of sale or purchase of, or the intent to sell or purchase, fish or wildlife or plants.

(b) Application of customs laws

All provisions of law relating to the seizure, forfeiture, and condemnation of property for violation of the customs laws, the disposition of such property or the proceeds from the sale thereof, and the remission or mitigation of such forfeiture, shall apply to the seizures and forfeitures incurred, or alleged to have been incurred, under the provisions of this chapter, insofar as such provisions of law are applicable and not inconsistent with the provisions of this chapter, except that all powers, rights, and duties conferred or imposed by the customs laws upon any officer or employee of the Treasury Department may, for the purposes of this chapter, also be exercised or performed by the Secretary or by such persons as he may designate: *Provided*, That any warrant for search or seizure shall be issued in accordance with rule 41 of the Federal Rules of Criminal Procedure.

(c) Storage cost

Any person convicted of an offense, or assessed a civil penalty, under section 3373 of this title shall be liable for the costs incurred in the storage, care, and maintenance of any fish or wildlife or plant seized in connection with the violation concerned.

(d) Civil forfeitures

Civil forfeitures under this section shall be governed by the provisions of chapter 46 of Title 18.

CREDIT(S)

(Pub.L. 97-79, § 5, Nov. 16, 1981, 95 Stat. 1076; Pub.L. 110-234, Title VIII, § 8204(d), May 22, 2008, 122 Stat. 1294; Pub.L. 110-246, Title VIII, § 8204(d), June 18, 2008, 122 Stat. 2056.)

§ 3375. ENFORCEMENT

(a) In general
   The provisions of this chapter and any regulations issued pursuant thereto shall be enforced by the Secretary, the Secretary of Transportation, or the Secretary of the Treasury. Such Secretary may utilize by agreement, with or without reimbursement, the personnel, services, and facilities of any other Federal agency or any State agency or Indian tribe for purposes of enforcing this chapter.

(b) Powers
   Any person authorized under subsection (a) of this section to enforce this chapter may carry firearms; may, when enforcing this chapter, make an arrest without a warrant, in accordance with any guidelines which may be issued by the Attorney General, for any offense under the laws of the United States committed in the person's presence, or for the commission of any felony under the laws of the United States, if the person has reasonable grounds to believe that the person to be arrested has committed or is committing a felony; may search and seize, with or without a warrant, in accordance with any guidelines which may be issued by the Attorney General; [FN1] Provided, That an arrest for a felony violation of this chapter that is not committed in the presence or view of any such person and that involves only the transportation, acquisition, receipt, purchase, or sale of fish or wildlife or plants taken or possessed in violation of any law or regulation of any State shall require a warrant; may make an arrest without a warrant for a misdemeanor violation of this chapter if he has reasonable grounds to believe that the person to be arrested is committing a violation in his presence or view; and may execute and serve any subpena, arrest warrant, search warrant issued in accordance with rule 41 of the Federal Rules of Criminal Procedure, or other warrant of civil or criminal process issued by any officer or court of competent jurisdiction for enforcement of this chapter. Any person so authorized, in coordination with the Secretary of the Treasury, may detain for inspection and inspect any vessel, vehicle, aircraft, or other conveyance or any package, crate, or other container, including its contents, upon the arrival of such conveyance or container in the United States or the customs waters of the United States from any point outside the United States or such customs waters, or, if such conveyance or container is being used for exportation purposes, prior to departure from the United States or the customs waters of the United States. Such person may also inspect and demand the production of any documents and permits required by the country of natal origin, birth, or reexport of the fish or wildlife. Any fish, wildlife, plant, property, or item seized shall be held by any person authorized by the Secretary pending disposition of civil or criminal proceedings, or the institution of an action in rem for forfeiture of such fish, wildlife, plants, property, or item pursuant to section 3374 of this title; except that the Secretary may, in lieu of holding such fish, wildlife, plant, property, or item, permit the owner or consignee to post a bond or other surety satisfactory to the Secretary.

(c) Jurisdiction of district courts
   The several district courts of the United States, including the courts enumerated in section 460 of Title 28, shall have jurisdiction over any actions arising under this chapter. The venue provisions of Title 18 and Title 28 shall apply to any actions arising under this chapter. The judges of the district courts of the United States and the United States magistrate judges may, within their respective jurisdictions, upon

proper oath or affirmation showing probable cause, issue such warrants or other process as may be required for enforcement of this chapter and any regulations issued thereunder.

(d) Rewards and incidental expenses

Beginning in fiscal year 1983, the Secretary or the Secretary of the Treasury shall pay, from sums received as penalties, fines, or forfeitures of property for any violation of this chapter or any regulation issued hereunder (1) a reward to any person who furnishes information which leads to an arrest, a criminal conviction, civil penalty assessment, or forfeiture of property for any violation of this chapter or any regulation issued hereunder, and (2) the reasonable and necessary costs incurred by any person in providing temporary care for any fish, wildlife, or plant pending the disposition of any civil or criminal proceeding alleging a violation of this chapter with respect to that fish, wildlife, or plant. The amount of the reward, if any, is to be designated by the Secretary or the Secretary of the Treasury, as appropriate. Any officer or employee of the United States or any State or local government who furnishes information or renders service in the performance of his official duties is ineligible for payment under this subsection.

CREDIT(S)

(Pub.L. 97-79, § 6, Nov. 16, 1981, 95 Stat. 1077; Pub.L. 98-327, § 4, June 25, 1984, 98 Stat. 271; Pub.L. 100-653, Title I,§ 104, Nov. 14, 1988, 102 Stat. 3826; Pub.L. 101-650, Title III, § 321, Dec. 1, 1990, 104 Stat. 5117.)

§ 3376. ADMINISTRATION

(a) Regulations

(1) The Secretary, after consultation with the Secretary of the Treasury, is authorized to issue such regulations, except as provided in paragraph (2), as may be necessary to carry out the provisions of sections 3372(f), 3373, and 3374 of this title.

(2) The Secretaries of the Interior and Commerce shall jointly promulgate specific regulations to implement the provisions of section 3372(b) of this title for the marking and labeling of containers or packages containing fish or wildlife. These regulations shall be in accordance with existing commercial practices.

(b) Contract authority

Beginning in fiscal year 1983, to the extent and in the amounts provided in advance in appropriations Acts, the Secretary may enter into such contracts, leases, cooperative agreements, or other transactions with any Federal or State agency, Indian tribe, public or private institution, or other person, as may be necessary to carry out the purposes of this chapter.

(c) Clarification of exclusions from definition of plant

The Secretary of Agriculture and the Secretary of the Interior, after consultation with the appropriate agencies, shall jointly promulgate regulations to define the terms used in section 3371(f)(2)(A) of this title for the purposes of enforcement under this chapter.

CREDIT(S)

(Pub.L. 97-79, § 7, Nov. 16, 1981, 95 Stat. 1078; Pub.L. 110-234, Title VIII, § 8204(e), May 22, 2008, 122 Stat. 1294; Pub.L. 110-246, Title VIII, § 8204(e), June 18, 2008, 122 Stat. 2056.)

§ 3377. EXCEPTIONS

(a) Activities regulated by plan under Magnuson-Stevens Fishery Conservation and Management Act

The provisions of paragraph (1) of section 3372(a) of this title shall not apply to any activity regulated by a fishery management plan in effect under the Magnuson-Stevens Fishery Conservation and Management Act (16 U.S.C. 1801 et seq.).

(b) Activities regulated by Tuna Convention Acts; harvesting of highly migratory species taken on high seas

The provisions of paragraphs (1), (2) (A), and (3) (A) of section 3372(a) of this title shall not apply to—

(1) any activity regulated by the Tuna Conventions Act of 1950 (16 U.S.C. 951-961) or the Atlantic Tunas Convention Act of 1975 (16 U.S.C. 971-971(h)); or

(2) any activity involving the harvesting of highly migratory species (as defined in paragraph (14) of section 3 of the Magnuson-Stevens Fishery Conservation and Management Act [16 U.S.C.A. § 1802(14)]) taken on the high seas (as defined in paragraph (13) of such section 3) if such species are taken in violation of the laws of a foreign nation and the United States does not recognize the jurisdiction of the foreign nation over such species.

(c) Interstate shipment or transshipment through Indian country of fish, wildlife, or plants for legal purposes

The provisions of paragraph (2) of section 3372(a) of this title shall not apply to the interstate shipment or transshipment through Indian country as defined in section 1151 of Title 18 or a State of any fish or wildlife or plant legally taken if the shipment is en route to a State in which the fish or wildlife or plant may be legally possessed.

CREDIT(S)

(Pub.L. 97-79, § 8, Nov. 16, 1981, 95 Stat. 1078; Pub.L. 104-208, Div. A, Title I, § 101(a) [Title II, § 211(b)], Sept. 30, 1996, 110 Stat. 3009-41.)

§ 3378. MISCELLANEOUS PROVISIONS

(a) Effect on powers of States

Nothing in this chapter shall be construed to prevent the several States or Indian tribes from making or enforcing laws or regulations not inconsistent with the provisions of this chapter.

(b) Repeals

The following provisions of law are repealed:

(1) The Act of May 20, 1926 (commonly known as the Black Bass Act; 16 U.S.C. 851-856).

(2) Section 667e of this title and sections 43 and 44 of Title 18 (commonly known as provisions of the Lacey Act).

(3) Sections 3054 and 3112 of Title 18.

(c) Disclaimers

Nothing in this chapter shall be construed as—

(1) repealing, superseding, or modifying any provision of Federal law other than those specified in subsection (b) of this section;

(2) repealing, superseding, or modifying any right, privilege, or immunity granted, reserved, or established pursuant to treaty, statute, or executive order pertaining to any Indian tribe, band, or community; or

(3) enlarging or diminishing the authority of any State or Indian tribe to regulate the activities of persons within Indianreservations.

(d) Travel and transportation expenses

The Secretary of the Interior is authorized to pay from agency appropriations the travel expense of newly appointed special agents of the United States Fish and Wildlife Service and the transportation expense of household goods and personal effects from place of residence at time of selection to first duty station to the extent authorized by section 5724 of Title 5 for all such special agents appointed after January 1, 1977.

(e) Interior appropriations budget proposal

The Secretary shall identify the funds utilized to enforce this chapter and any regulations thereto as a specific appropriations item in the Department of the Interior appropriations budget proposal to the Congress.

CREDIT(S)

(Pub.L. 97-79, § 9(a)-(c), (g), (h), Nov. 16, 1981, 95 Stat. 1079, 1080.)

Bibliography

Abrar, Ahmed. *Live Bird Trade in Northern India*. TRAFFIC India, 1997.

Agustini, Rina, Iola Leal Riesco, and Ridzki Rinanto Sigit. "Finding Solutions to Illegal Logging: Civil Society and the FLEGT Support Project." 2005. http://www.fern.org/sites/fern.org/files/media/documents/document_1287-1288.pdf (accessed November 1, 2010).

Albini, Joseph, R. E. Rogers, Victor Shabalin, Valery Kutushev, Vladimir Moiseev, and Julie Anderson. "Russian Organized Crime: Its History, Structure, and Function." *Journal of Contemporary Criminal Justice* 11, no. 4 (1995): 214–43.

Alcantara, Odette. "Segregate, Compost, Recycle: Tutorials for Slow Learners." http://www.journal.com.ph/index.php?issue=209-07-08&sec=14&aid=97684 (accessed 2009).

Anyinam, CA. "Transboundary Movements of Hazardous Wastes: The Case of Toxic Waste Dumping in Africa." *International Journal of Health Service* 21, no. 4 (1991): 759–77.

Arborvitae: The IUCN/WWF Forest Conservation Newsletter 32 (December 2006). http://www.illegal-logging.info/uploads/arborvitae_32.pdf (accessed November 1, 2010).

Ashwell, David, and Naomi Walston. *An Overview of the Use and Trade of Plants and Animals in Traditional Medicine Systems in Cambodia*. TRAFFIC Southeast Asia, 2008.

Asia Pulp & Paper (APP) Threatens Bukit Tigapuluh Landscape: Report of Investigation Findings. WWF, Indonesia, 2008. http://www.illegal-logging.info/uploads/APPinvestigationjan2008.pdf

Auliya, Mark. *Hot Trade in Cool Creatures: A Review of the Live Reptile Trade in the European Union in the 1990s with a Focus on Germany*. TRAFFIC Europe, 2003.

Barber, C. V., and R. V. Pratt. "Poison and Profits: Cyanide Fishing in the Indo-Pacific." *Environment* 40 (1998): 5–34.

Berlchoudt, Karin. *Focus on EU Enlargement and the Wildlife Trade: A Review of CITES Implementation in Candidate Countries.* TRAFFIC Europe, 2002.

Berry, Jessica. "Armed Gangs Threaten World Caviar Stocks." *Sunday Telegraph,* July 18, 1999. http://www.highbeam.com/doc/1P2-19129915.html (accessed October 6, 2010).

Block, Alan A. *The Business of Crime.* Boulder, CO: Westview, 1991.

Block, Alan A., and Frank R. Scarpitti. *Poisoning for Profit: The Mafia and Toxic Waste Disposal in America.* New York: William Morrow, 1985.

"Bluefin Tuna, Polar Bear Denied Export Protections." *Tribune Review,* March 19, 2010, A2.

Brack, Duncan. "Controlling Illegal Logging: Lessons from the US Lacey Act." Chatham House: Energy, Environment and Development Programme. 2007. http://www.euflegt.efi.int/uploads/16LaceyActbp0702.pdf (accessed November 1, 2010).

Brack, Duncan. "The Growth and Control of International Environmental Crime." *Environmental Health Perspectives,* 112, no.2 (2004): 80–81.

Brack, Duncan, Kevin Gray, and Gavin Hayman. *Controlling the International Trade in Illegally Logged Timber and Wood Products.* A study prepared for the UK Department for International Development. London: Royal Institute of International Affairs, 2002.

"Britain's Filthy Garbage Causes Stink in Ports." *Tribune Review,* July 17, 2009, A2.

Broken Promises: How World Bank Group Policies Fail to Protect Forests and Forest Peoples' Rights. The Rainforest Foundation, CDM Watch, Global Witness, SinksWatch, Forest Peoples Programme, Environmental Defense, World Rainforest Movement, Down to Earth, 2005. http://www.forestpeoples.org/sites/fpp/files/publication/2010/08/wbbrokenpromisesapr05eng.pdf (accessed November 1, 2010).

Bulldozing Progress: Human Rights Abuses and Corruption in Papua New Guinea's

Large Scale Logging Industry. Australian Conservation Foundation and the Center for Environmental Law and Community Rights. 2006. http://www.acfonline.org.au/uploads/res/res_acf-celcor_full.pdf (accessed October 4, 2010).

"California Advances Grocery Store Plastic Bag Ban." *Tribune* Review, June 4, 2010, A4.

Carter, Timothy S. "The Failure of Environmental Regulation in New York: The Role of Cooptation, Corruption and Cooperative Enforcement Approach." *Crime, Law and Social Change* 26, no. 1 (1996): 27–52.

"Caviar Hunters Pushing Sturgeon to Extinction's Edge." Illegal-Fishing.info, March 18, 2010. http://www.illegal-fishing.Info/item_single.php?item=news&item_id=4624&approach_id= (accessed September 17, 2010).

"China Set to Curb Foreign Waste Imports." China Daily, January 24, 2007. http://www.chinadaily.com.cn/china/2007-01/24content791722.htm (accessed September 19, 2010).

Ciffullo, Frank J., Sharon L. Cardash, and Gordon N. Lederman. *Combating Chemical, Biological, Radiological and Nuclear Terrorism: A Comprehensive Strategy*. Center for Strategic and International Security Studies. Washington, DC: CSIS Press, 2002.

Clapp, Jennifer. "Africa, NGOs, and the International Toxic Waste Trade." *Journal of Environment and Development* 3, no. 2 (1994): 17–46.

Clapp, Jennifer. "The Illicit Trade in Hazardous Wastes and CFCs: International Responses to Environmental Bads." *Trends in Organized Crime* 3, no. 2 (1997): 14–18.

Clark, Colin W., Gordon R. Munro, and Ussif Rashid Sumalia. "Subsidies, Buybacks, and Sustainable Fisheries." *Journal of Environmental Economics and Management* 50, no. 1 (2005): 47–58.

Colombo, Francesca. "Animal Trafficking—A Cruel Billion Dollar Business." *Inter Press Service*, September 6, 2006. http://www.commondreams.org/headlines03/0906-06.htm (accessed September 24, 2010).

Colombo, Francesca. "Mafia Dominates Garbage Industry." Tierramérica, 2003. http://www.tierramerica.net/2003/0623/iarticulo.shtml (accessed October 6, 2010).

Comptroller General's Report to the Subcommittee on Investigations and Oversight, Committee on Public Works and Transportation, House of Representatives. *Illegal Disposal of Hazardous Waste: Difficult to Detect or Deter*. Washington, DC: U.S. General Accounting Office, 1985.

Cook, Dee, Martin Roberts, and Jason Lowther. *The International Wildlife Trade and Organised Crime: A Review of the Evidence and the Role of the UK*. Regional Research Institute: University of Wolverhampton, 2002.

Cooke, Andrew, and Wendy Chapple. "Merger Activity in the Waste Disposal Industry: The Impact and the Implications for the Environmental Protection Act." *Applied Economics* 32, no. 6 (2000): 749–55.

Czarnomski, Sarah, Barry Webb, and Alan Holmes. *IMPEL-TFS Threat Assessment Project: The Illegal Shipment of Waste Among IMPEL Member States*. Jill Dando Institute of Crime Science, 2005.

"Data on the Black Market in Illegal Logging." Havocscope. http://www.havoscope.com/trafficking/logging.htm (accessed October 7, 2010).

"Data on the Black Market in Wildlife and Animal Smuggling." Havocscope. http:/www./havoscope.com/trafficking/wildlife.htm (accessed October 7, 2010).

Davies, Ben. *Black Market: Inside the Endangered Species Trade in Asia*. San Rafael, CA: Earth Aware Editions, 2005.

"Demand Has Elephants on Precipice." *Tribune Review*, June 16, 2010, A15.

Dorn, Nicholas, Stijn Van Daele, and Tom Vander Beken. "Reducing Vulnerabilities to Crime of the European Waste Management Industry: The Research Base and the Prospects for Policy." *European Journal of Crime, Criminal Law, and Criminal Justice* 15, no. 1 (2007): 23–36.

Dronova, Natalia, and Vassily Spiridonov. *Illegal, Unreported, and Unregulated Pacific Salmon Fishing in Kamchatka*. WWF-Russia and TRAFFIC Europe-Russia, 2008.

Dykstra, Dennis P., George Kuru, Rodney Taylor, Ruth Nussbaum, William B. Magrath, and Jane Story. "Technologies for Wood Tracking: Verification and Monitoring the Chain of Custody and Legal Compliance in the Timber Industry." Environmental and Social Development East Asia and Pacific Region Discussion Paper. World Bank, March 18, 2009. http://www.illegal-logging.info/item_single.php ?it_id=30&it=document (accessed October 6, 2010).

Electronic Waste and Organized Crime: Assessing the Links. Phase II Report for the INTERPOL Pollution Crime Working Group. May 2009. http://www.interpol.int/Public/ICPO/FactSheets/WasteReport.pdf (accessed September 19, 2010).

"Enviro Brief: Degradable Plastic Waste Sacks." RedOrbit, January 17, 2006. http://www.redorbit.com/news/science/359524/enviro_brief _degradable_plastic_waste_sacks/index.html (accessed October 4, 2010).

"Environmental Effects of Illicit Crop Cultivation." *Trends in Organized Crime* 3, no. 2 (1997): 11–14.

"The Environmental Effects of Illicit Crop Cultivation." *The United Nations International Drug Control Programme, World Drug Report*. New York: Oxford University Press, 1997.

Exporting Destruction: Export Credits, Illegal Logging and Deforestation. FERN. 2008. http://www.fern.org/media/documents/document_4155_4160.pdf (accessed October 4, 2010).

Fagan, Chris, and Diego Shoobridge. *The Race for Peru's Last Mahogany Trees: Illegal Logging and the Alto Purus National Park*. Round River Conservation Studies, 2007.

Faiola, Anthony. "Smuggling's Wild Side in Brazil: Animal Trafficking Sucks the Life from the Amazon Rain Forest." *Washington Post*, December 9, 2001, A38. http://www.latinamericanstudies.org/brazil/ smuggling.htm (accessed November 1, 2010).

"Four Algerians, Five Turks Jailed for Illegal Fishing." Illegal-Fishing.info, April 5, 2010. http://www.illegal-fishing.info/item_single.php?item=news &item=news&item_id=4641&approach_id= (accessed September 17, 2010).

Freemantle, Brian. *The Octopus: Europe in the Grip of Organized Crime*. London: Orion, 1995.

Galster, Steven R., S. F. LaBudde, and C. Stark. *Crimes against Nature: Organized Crime and the Illegal Wildlife Trade*. The Endangered Species Project, 2004.

Glastra, Rob. *Cut and Run: Illegal Logging and Timber Trade in the Tropics*. Ottawa, Canada: IDRC Books, 1999.

Gobbi, Jose, Debra Rose, Gina De Ferrari, and Leonora Sheeline. *Parrot Smuggling across the Texas-Mexico Border*. TRAFFIC USA, 1996.

Godoy, Julio. "Environment-France: Dismantling End-of-Life Ships Requires Global Answers." *Inter Press Service*, June 19, 2006. http://www.ipsnews .net/news.asp?idnews=33675 (accessed September 17, 2010).

Goodman, Peter S., and Peter Finn. "Corruption Stains the Timber Trade: Forests Destroyed China's Race to Feed Global Wood-Processing

Industry." *Washington Post Foreign Service*, April 1, 2007. http://www.washingtonpost.com/wp-dyn/content/article/2007/03/31/AR2007033101287.html (accessed November 1, 2010).

"Government Officials Arrested in Logging Bust." Illegal-Logging.info, April 5, 2010. http://www.illegal-logging.info/item_single.php?it_id=4321&it=news (accessed September 17, 2010).

Green, Alan. *Animal Underworld: Inside America's Black Market for Rare and Exotic Species*. New York: Public Affairs, 1999.

"GreenShift Announces New Commercial Appliance for Reduction of Plastic Waste." RedOrbit, July 18, 2005. http://www.redorbit.com/news/technology/177818/greenshift_announces_new_commerical_appliance_for_reduction_of_plastic_waste/index.html (accessed October 4, 2010).

Grossman, Elizabeth. *High Tech Trash: Digital Devices, Hidden Toxics and Human Health*. Washington: Island Press/Shearwater Books, 2006.

Grumbles, Benjamin H. "EPA's Marine Debris Program: Taking Action against Trash." 2008. RedOrbit, http://www.redorbit.com/news/science/1252662/epas_marine_debris_program_taking_action_against_trash/index.html (accessed October 6, 2010).

"Guangzhou Customs Intercepts 237 Tons of Smuggled Trash." China.org.cn, 2007. http://japanese.china.org.cn/english/SO-e/27771.htm (accessed September 19, 2010).

Hayman, Gavin, and Duncan Brack. *International Environmental Crime: The Nature and Control of Black Markets*. London: The Royal Institute of International Affairs, 2002.

Hearings held by the Subcommittee on Oversight and Investigations, Committee on Interstate and Foreign Commerce, House of Representatives. *Organized Crime and Hazardous Waste Disposal*. Washington, DC: U.S. Government Printing Office, 1980.

Henry, Leigh. *A Tale of Two Cities: A Comparative Study of Traditional Chinese Medicine Markets in San Francisco and New York City*. TRAFFIC North America. Washington DC: World Wildlife Fund, 2004.

Hewitt, James. *Failing the Forests: Europe's Illegal Timber Trade*. WWF-UK, 2005.

Hilz, Christoph. *The International Toxic Waste Trade*. New York: Van Nostrand Reinhold, 1992.

Hin Keong, Chen. *The Role of CITES in Combating Illegal Logging: Current and Potential*. Cambridge: TRAFFIC International, 2006.

Holden, Jane. *By Hook or by Crook: a Reference Manual on the Illegal Wildlife Trade and Prosecutions in the United Kingdom*. United Kingdom: The Royal Society for the Protection of Birds, WWF-UK, TRAFFIC International, 1998.

Hongfa, Xu, and Craig Kirkpatrick. *The State of Wildlife Trade in China: Information on the Trade in Wild Animals and Plants in China 2007*. TRAFFIC East Asia China Programme, 2007.

Illegal-Fishing.info. http://www.illegal-fishing.info

"Illegal Logging Financing Taliban Attacks on U.S. Troops." Illegal-Logging.info, April 16, 2010. http://www.illegal-logging.info/item _single.php?it_id=4353&it=news (accessed September 17, 2010).

Illegal Logging, Governance, and Trade: 2005 Joint NGO Conference. FERN/Greenpeace/WWF, 2005. http://www.fern.org/sites/fern.org/files/media/documents/document_1650_1659.pdf (accessed November 1, 2010).

The Illegal Trade in Wild Birds for Food through Southeast and Central Europe. TRAFFIC, 2008.

"Illicit Trafficking and Other Unauthorized Activities Involving Nuclear and Radioactive Materials." http://www.iaea.org/NewsCenter/Features/RadSources/PDF/fact_figures2007.pdf (accessed November 1, 2010).

"India's E-Trash Safety Push Raises Concern." *Tribune Review*, June 13, 2010, A11.

International Atomic Energy Agency. *IAEA Illicit Trafficking Database, Fact Sheet.* 2008.

"International Crime Threat Assessment." *Trends in Organized Crime* 5, no. 4 (2000): 56–59.

Jacobs, James, Christopher Panarella, and Jay Worthington. *Busting the Mob: U.S. v. Cosa Nostra.* New York: New York University Press, 1994.

Janiabe, Sarah. "Luxury Store Owner Convicted for Wildlife Trafficking." World Wildlife Fund, August 27, 2007. http://www.worldwildlife.org/who/media/press/2007/WWFPresitem987.html (accessed October 7, 2010).

Johnson, Andrea. "U.S. Lacey Act: Respecting the Laws of Trade Partners." *Jakarta Post*, March 5, 2009. http://www.thejakartapost.com/news/2009/03/05/us-lacey-act-respecting-laws-trade-partners.html (accessed November 1, 2010).

"Just the Facts." *Gazette* (Royal Canadian Mounted Police) 66, no. 3 (2004).

Kecse-Nagy, Katalin, Dorottya Papp, Amelie Knapp, and Stephanie Von Meibom. *Wildlife Trade in Central and Eastern Europe: A Review of CITES Implementation in 15 Countries.* TRAFFIC Europe, 2006.

Lack, Mary, and Glenn Sant. *Illegal, Unreported, and Unregulated Shark Catch: A Review of Current Knowledge and Action.* TRAFFIC International, 2008.

Lack, Mary. *Trends in Global Shark Catch and Recent Developments in Management.* TRAFFIC International, 2009.

Lee, Rensselaer. "Recent Trends in Nuclear Smuggling." *Transnational Organized Crime* 2, no. 4 (1996): 109–21.

Legambiente, Gruppo Abele-Nomos. *The Illegal Trafficking in Hazardous Waste in Italy and Spain: Final Report.* 2003. http://www.organized-crime.de/revgru03.htm (accessed November 1, 2010).

Levy, Adrian, and Cathy Scott-Clark. "Poaching for bin Laden."*Guardian*, May 5, 2007. http://www.guardian.co.uk/world/2007/may/05/terrorism.animalwelfare (accessed November 1, 2010).

Lyapustin, Sergey N., Alexey L. Vaisman, and Pavel V. Fomenko. *Wildlife Trade in the Russian Far East: An Overview.* TRAFFIC Europe-Russia, 2007.

Macabrey, Jean-Marie. "Deforestation: E.U. Committee Backs Broad New Penalties on Illegal Timber Trade." E&E Publishing, February 18, 2009. http://www.eenews.net/public/climatewire/2009/02/18/3 (accessed October 4, 2010).

"Mafia Sank Boat with Radioactive Waste: Official." Nuclear Power Daily, September 14, 2009, http://www.nuclearpowerdaily.com/reports/Mafia_sank_boat_with_radioactive_waste_official_999.html (accessed September 19, 2010).

"Massacre in Peru: A Dispatch on the Bloody Conflict." Banderas News, June 9, 2009. http://www.banderasnews.com/0906/edat-bloodyconflict.htm (accessed October 4, 2010).

Massari, Monica. "Ecomafias and Waste Entrepreneurs in the Italian Market." Paper presented at the 6th Cross-border Crime Colloquium, Berlin, Germany, 2004.

Massari, Monica, and Paola Monzini. "Dirty Business in Italy: a Case Study of Trafficking in Hazardous Waste." *Global Crime* 6. no. 3–4 (2004): 285–304.

McManus, John W., Rodolfo B. Reyes Jr., and Cleto L. Nanola Jr. "Effects of Some Destructive Fishing Practices on Coral Cover and Potential Rates of Recovery." *Environmental Management* 21, no. 1 (1997): 69–78.

Milliken, Tom, R. W. Burn, and L. Sangalakula. *The Elephant Trade Information System (ETIS) and the Illicit Trade in Ivory.* TRAFFIC East/Southern Africa, 2009.

Milliken, Tom, Alistair Pole, and Abias Huongo. "No Peace for Elephants: Unregulated Domestic Ivory Markets in Angola and Mozambique." TRAFFIC East/Southern Africa, 2006.

"Mob Expert Says Naples Garbage Fix Only Temporary." FoxNews.com, January 9, 2008. http://www.foxnews.com/story/0,2933,321410,00.html (accessed September 19, 2010).

Mobile Toxic Waste: Recent Findings on the Toxicity of End-of-Life Cell Phones. A Report by Basel Action Network (BAN), 2004.

Montlake, Simon. "Indonesia Battles Illegal Timber Trade." Special to the *Christian Science Monitor*, February 27, 2002. http://www.csmonitor.com/2002/0227/p07s01-woap.html (accessed October 4, 2010).

Moyers, Bill D. *The Global Dumping Ground.* Cambridge: Lutterworth Press, 1991.

Nash, Stephen V. *Sold for a Song: The Trade in Southeast Asian non-CITES Birds.* TRAFFIC Southeast Asia, 1993.

Ng, Julia, and Nemora. *Tiger Trade Revisited in Sumatra, Indonesia.* TRAFFIC South East Asia, 2007.

Nijman, Vincent. *In Full Swing: An Assessment of Trade in Orang-utans and Gibbons on Java and Bali, Indonesia.* TRAFFIC Southeast Asia, 2005.

Nowell, Kristin, and Xu Ling. *Taming the Tiger: China's Markets for Wild and Captive Tiger Products since the 1993 Domestic Trade Ban.* TRAFFIC East Asia, 2007.

"Nuclear News: EPA to Rebuild Uranium-Contaminated Navajo Homes." June 15, 2009. Weblog.greenpeace.org/nuclear- reaction/2009/06/nuclear_news_epa_to_rebuild_ur.html (accessed September 19, 2010).

"Ocean Protection Council to Fight Marine Debris, Fund Low-Interest Loans to Fishing Businesses and Communities." Business Wire, February 7, 2007. http://www.thefreelibrary.com/Ocean+Protection+Council+to+Fight+Marine+Debris,+Fund+Low-Interest...-a0158978406 (accessed November 1, 2010).

"Organized Crime and the Environment." *Trends in Organized Crime* 3, no. 2 (1997): 4.

"Organized Crime Fuels Illegal Ivory Surge in Africa." http://www.panda.org (accessed September 24, 2010).

Paul, Katie. "Exporting Responsibility: Shipbreaking in South Asia: International Trade in Hazardous Waste." *Environmental Policy and Law* 34 (2004): 73–78.

Pellow, David Naguib. *Resisting Global Toxics: Transnational Movements for Environmental Justice.* Cambridge, MA: MIT Press, 2007.

Pfeiffer, Tom. "Whaling Moratorium Talks Break Down." June 23, 2010. http://uk.reuters.com/article/idUKTRE65M26P20100623 (accessed November 1, 2010).

Pirates and Profiteers: How Pirate Fishing Fleets are Robbing People and Oceans. London: Environmental Justice Foundation, 2005.

Pomeroy, Robin. "Naples Garbage Is Mafia Gold." Reuters, January 9, 2008. http://www.reuters.com/article/idUSL083057720080109 (accessed September 19, 2010).

"Pot Farms Wreak Havoc in Sequoia National Park." 2005. http://seattletimes.nwsource.com/html/nationworld/2002445217_marijuana19.html (accessed October 7, 2010),

Pratt, V. R. "The Growing Threat of Cyanide Fishing in the Asia Pacific Region and the Emerging Strategies to Combat It." *Coastal Management in Tropical Asia* 5 (1996): 9–11.

"Public Benefit Company Grows in Response to Concern about Plastic Bag Waste." RedOrbit, March 14, 2006. http://www.redorbit.com/news/science/427251/public_benefit_company_grows_in_response_to_concern_about_plastic/index.html (accessed November 1, 2010).

Pye-Smyth, Charlie. *Logging in the Wild East: China and the Forest Crisis in the Russian Far East.* TRAFFIC, 2006.

Rebovich, Donald J. *Dangerous Ground: The World of Hazardous Waste Crime.* New Brunswick, NJ: Transaction Publishing, 1992.

"Recycling of PVC and Mixed Plastic Waste: Bringing the Recycling Process from a Costly Nuisance to a Profitable Industry." RedOrbit, October 12, 2005. http://www.redorbit.com/news/science/268954/recycling_of_pvc_and_mixed_plastic_waste_bringing_the/ (accessed October 7, 2010).

Reuter, Peter, Jonathan Rubenstein, and Simon Wynn. *Racketeering in Legitimate Industries: Two Case Studies*. Washington, DC: National Institute of Justice, 1983.

Review of the Impacts of Illegal, Unreported and Unregulated Fishing on Developing Countries, Final Report. London: Marine Resources Assessment Group Ltd., 2005.

"Rhino Poaching Soars." News24.com, March 21, 2010. http://www.news 24.com/SciTech/News/Rhino-poaching-soars-20100321 (accessed November 1, 2010).

Rhodes, William M., Elizabeth P. Allen, and Myfanwy Callahan. *Illegal Logging: A Market-Based Analysis of Trafficking in Illegal Timber*. Cambridge, MA: Ibt Associates Inc., 2006.

Richards, M., A. Wells, F. Del Gatto, A. Contreras-Hermosilla, and D. Pommier. "Impacts of Illegality and Barriers to Legality: A Diagnostic Analysis of Illegal Logging in Honduras and Nicaragua." *International Forestry Review* 5, no. 3 (2003): 282–92.

"Salamander Protected from Trade." *Tribune Review*, March 22, 2010, A3.

Saviano, Roberto. *Gomorrah: A Personal Journey into the Violent International Empire of Naples' Organized Crime System*. New York: Farrar, Straus, and Giroux, 2007.

Schafer, Kristin S. "One More Failed U.S. Environmental Policy." Foreign Policy in Focus, August 31, 2006, http://www.fpif.org/reports/one _more_failed_us_environmental_policy (accessed September 19, 2010).

Schmidt, Charles W. "Environmental Crimes: Profiting at the Earth's Expense." *Environmental Health Perspectives*, 112, no. 2 (2004).

"2nd Garbage Patch Discovered." *Tribune Review*, April 16, 2010, A3.

Secretariat of the Basel Convention. "The Basel Convention at a Glance." http://www.basel.int/convention/bc_glance.pdf (accessed September 19, 2010).

Shivers, C. J. "Corruption Endangers a Treasure of the Caspian." *New York Times*, November 28, 2005. http://www.nytimes.com/2005/11/28/ international/asia/28sturgeon.html (accessed October 6, 2010).

"Smugglers Find Treasure in Trash." People's Daily Online, December 29, 2007, http://english.peopledaily.com.cn/90001/90776/6329761.html (accessed September 19, 2010).

Stiles, Daniel. *The Elephant and Ivory Trade in Thailand*. TRAFFIC Southeast Asia, 2009.

Strohm, Laura A. "The Environmental Politics of the International Waste Trade." *Journal of Environment and Development* 2, no. 2 (1993).

Study and Analysis of the Status of IUU Fishing in the SADC Region and an Estimate of the Economic, Social and Biological Impacts, Volume 2, Main Report. London: Marine Resources Assessment Group Ltd., 2008.

Sumaila, Ussif Rashid, Morten D. Skogen, David Boyer, and Stein Ivansteinshamn. *Namibia's Fisheries: Ecological, Economic and Social Aspects*. Delft, the Netherlands: Eburon Academic Publishers, 2004.

"Switching Channels: Wildlife Trade Routes into Europe and the UK."
    WWF/TRAFFIC, December 2002. http://www.wwf.org.uk/filelibrary/
    pdf/switchingchannels.pdf (accessed November 1, 2010).

Szasz, Andrew. "Corporations, Organized Crime, and the Disposal of
    Hazardous Waste: an Examination of the Making of a Criminogenic
    Regulatory Structure." *Criminology* 24, no. 1 (1986): 1–27.

"Tainted Water Kills 6 Members of Tribe." *Tribune Review*, April 16, 2010,
    A2.

"Tangled Global Web Cited in Spill." *Tribune Review*, June 15, 2010, A1.

Thomson, Jamie, and Ramzy Kanaan. *Conflict Timber: Dimensions of the
    Problem in Asia and Africa*. Volume 1. Burlington, VT: ARD, Inc.,
    2004.

Tonetti, Robert. "Export of Used and Scrap Electronics: What You Need to
    Know." EPA Office of Solid Waste Presentation. http://www.epa.gov/
    epaoswer/hazwaste/recycle/ecycling/rules.htm (accessed 2009).

"Toxic Tech: Pulling the Plug on Dirty Electronics." Accessed 2009.
    www.greenpeace.org/raw/content/international/press/reports/toxic-tech
    -puling-the-plug-o.pdf.

Traffic: The Wildlife Trade Monitoring Network. http://www.traffic.org.

Trends in Organized Crime 3, no. 2 (1997): 23–24.

"Underwater Plastic Waste Threatens World's Food Chain." RedOrbit,
    March 27, 2008. http://www.redorbit.com/news/science/1314953/
    underwater_plastic_waste_threatens_worlds_food_chain/index.html
    (accessed October 7, 2010).

"United States Forms Global Coalition Against Wildlife Trafficking,"
    September 23, 2005. http://www.america.gov/st/washfile-english/2005/
    September/20050923155154lcnirellep0.4162409 (accessed November 1,
    2010).

U.S. Congress. House of Representatives. *Organized Crime Links to the Waste
    Disposal Industry*. 97th Cong., 1st sess. Washington, DC: U.S. Government
    Printing Office, 1981.

U.S. Congress. Senate. Permanent Subcommittee on Investigations of
    the Committee on Governmental Affairs, *Profile of Organized Crime:
    Mid-Atlantic Region*. 98th Cong., 1st sess. Washington, DC: U.S.
    Government Printing Office, 1983.

"Violent Mob Objects to Crackdown on Illegal Logging." *Tribune Review*,
    November 25, 2008, A2.

Wald, Matthew L. "Pentagon Puts Millions into Fuel for Plastics." RedOrbit,
    April 9, 2007. http://www.redorbit.com/news/science/896830/pentagon
    _puts_millions_into_fuel_from_plastic/index.html (accessed October 4,
    2010).

Webster, Donovan (1997). "The Looting and Smuggling and Fencing and
    Hoarding of Impossibly Precious, Feathered, and Scaly Wild Things."
    Trends in Organized Crime 3, no. 2 (1997): 9–10.

"What Is CITES?" http://www.cites.org/eng/disc/what.shtml (accessed
    2009).

What's Driving the Wildlife Trade? A Review of Expert Opinion on Economic and Social Drivers of the Wildlife Trade and Trade Control Efforts in Cambodia, Indonesia, Lao PDR, and Vietnam. East Asia and Pacific Region Sustainable Development Department. Washington, DC: World Bank, 2008.

"Wildlife Group Nixes Shark Protections." *Tribune Review*, March 24, 2010, A6.

Woessner, Paul N. "Chronology of Radioactive and Nuclear Materials Smuggling Incidents: July 1991–June 1997." *Transnational Organized Crime* 3 (1997): 114–209.

World Wildlife Fund. http://www.worldwildlife.org.

"WRM Bulletin 98" (January 9, 2005; edited March 18, 2009). http://www.illegal-logging.info/item_single.php?item=document&item_id=237&approach_id=15 (accessed October 4, 2010).

Index

About the Author

DONALD R. LIDDICK received his doctoral degree from Pennsylvania State University in 1995. Since then he has worked at the University of Pittsburgh at Greensburg, and now teaches criminology and criminal justice courses at Penn State–Fayette, the Eberly Campus. *Crimes Against Nature* is Dr. Liddick's sixth book. He lives with his sons in the Laurel Highlands region of Western Pennsylvania.